Timber Buildings and Sustainability

Edited by Giovanna Concu

Published in London, United Kingdom

IntechOpen

Supporting open minds since 2005

Timber Buildings and Sustainability
http://dx.doi.org/10.5772/intechopen.78428
Edited by Giovanna Concu

Contributors
Meng Gong, Ali M. Memari, Anthony Jellen, Cecilia Mark-Herbert, Elin Kvennefedt, Anders Roos, Hasan Tezcan Yildirim, Frederick Cubbage, Joshua Roll, Patrcio MacDonagh, George Hahn, Francesco Carbone, Rodolfo Picchio, Giovanna Concu

First published in London, United Kingdom, 2019 by IntechOpen
IntechOpen is the global imprint of INTECHOPEN LIMITED, registered in England and Wales, registration number: 11086078, 7th floor, 10 Lower Thames Street, London, EC3R 6AF, United Kingdom
Printed in Croatia

British Library Cataloguing-in-Publication Data
A catalogue record for this book is available from the British Library

Additional hard and PDF copies can be obtained from orders@intechopen.com

Timber Buildings and Sustainability
Edited by Giovanna Concu
p. cm.
Print ISBN 978-1-78923-881-5
Online ISBN 978-1-78923-882-2
eBook (PDF) ISBN 978-1-78985-843-3

For EU product safety concerns:
IN TECH d.o.o., Prolaz Marije Krucifikse Kozulić 3, 51000 Rijeka, Croatia, info@intechopen.com or visit our website at intechopen.com.

Meet the editor

Giovanna Concu is Assistant Professor of Structural Design at the University of Cagliari, Italy, where she teaches Timber and Masonry Structures and Laboratory of Structural Rehabilitation. She studied Structural Engineering and was awarded a PhD in Geoengineering. She has authored about eighty papers and is a reviewer for international journals and scientific associations. She has been involved in a number of research projects funded by regional, national and international public and private agencies. Her research interests include sustainability in buildings and constructions, timber engineering, and non-destructive testing. Currently she is dealing with the mechanical behavior of cross-laminated and glue-laminated timber elements made with low-graded timber and in the development of short supply chain and circular economy schemes that aim at the use of local materials for the building industry.

Contents

Preface XI

Chapter 1 1
Introductory Chapter: Timber and Sustainability in Construction
by Giovanna Concu

Chapter 2 7
Lumber-Based Mass Timber Products in Construction
by Meng Gong

Chapter 3 27
Structural Design of a Typical American Wood-Framed Single-Family Home
by Anthony C. Jellen and Ali M. Memari

Chapter 4 55
Communicating Added Value in Wooden Multistorey Construction
by Cecilia Mark-Herbert, Elin Kvennefeldt and Anders Roos

Chapter 5 69
Operative Machinery Costs Analysis within Forest Management Implementation Frame
by Francesco Carbone and Rodolfo Picchio

Chapter 6 101
Raw Material Demand-Supply and Policy Recommendations of Turkish Wood-Based Panel Industry
by Hasan Tezcan Yildirim

Chapter 7 115
Timber Harvesting Production, Costs, Innovation, and Capacity in the Southern Cone and the U.S. South
by Patricio Mac Donagh, Joshua Roll, George Hahn and Frederick Cubbage

Preface

The theme of environmental sustainability is highly topical and requires a rethinking of the consumer models to which we have been accustomed to date. All sectors of the productive world are affected by this change of paradigm, and the construction sector, which alone accounts for 40 percent of resource consumption and environmental pollution, plays an important role. Focusing attention on building materials, the trend is to aim for short chain and circular building schemes, pushing production and use of natural materials that are locally produced, recyclable and renewable. In this perspective, a great reassessment is reserved for timber, which offers all the aforementioned qualities in addition to being a resistant, comfortable and economical material. The further impulse to the spread of timber in buildings goes through different phases, such as the optimization of forest management, the implementation of an effective production chain, the information of operators about the opportunities related to the use of this material, and the indispensable phases of scientific research and technological development in order to guarantee best performance.

This book provides insight into the spread of the use of timber in the construction industry, presenting some thoughts on important aspects of the topic such as:

- construction systems that can be implemented by using timber
- the basic aspects of correct timber constructions design
- characteristics of the timber buildings market and the importance of its analysis
- the sustainability analysis of all phases of the production process
- the importance of the analysis of production costs

The themes are addressed both through general approaches and through the reference to specific case studies able to clearly illustrate the issues under consideration.

Timber is a natural building material. If used in building elements, it can play structural functions and aesthetic roles at the same time. The use of timber in buildings, which goes back to the oldest of times, is now experiencing a period of strong expansion in virtue of the material's sustainability. However, timber's use as an engineering material calls for constant development of theoretical and experimental research to properly respond to the issues involved. The book aims to contribute to knowledge in the application of timber in the building industry.

I wish to express my gratitude to all the authors and co-authors of the chapters for their interest and valuable contributions. I also wish to thank the publishing process manager and the editorial and production staff at IntechOpen for their assistance and professional support.

Giovanna Concu
Department of Civil and Environmental Engineering and Architecture,
University of Cagliari,
Cagliari, Italy

Chapter 1

Introductory Chapter: Timber and Sustainability in Construction

Giovanna Concu

1. Introduction

The buildings built in the second half of the last century and in the first decade of this century are characterized by the enormous consumption of energy and natural resources, so that the consumption of natural resources, the production of a considerable amount of waste, and the pollution of air and water are the largest undesirable effects related to the construction industry. Currently, the construction industry consumes annually an average of 40% of resources such as raw materials, water, and energy and produces about 40% of solid waste and pollution. In this context it becomes imperative for the building sector to move towards a sustainable dimension. This is the reason why in recent years a cultural model based on environmental sustainability has developed, involving all sectors of human activity and evidently the construction sector. Building activity is increasingly geared towards respecting, safeguarding, and enhancing the environment, through the design and use of materials and production and management processes based on biocompatibility, energy saving, and the green economy.

The concept of sustainable development was made explicit, for the first time, in a document presented by G.H. Brundtland in a meeting of the World Commission on Environment and Development [1]. The document, better known as the Brundtland Report, Our Common Future, found that the critical points and global problems of the environment are essentially due to the great poverty of the south of the world and to the unsustainable production and consumption patterns of the north of the world. The report therefore highlighted the need to implement a strategy capable of integrating the needs of development and the environment. This strategy was defined with the term *sustainable development*, whose precise definition was as follows: *development is sustainable if it meets the needs of the present without compromising the ability of future generations to meet their own needs*. This definition contained a new concept relating to sustainable development, able to reconcile aspects such as expectations of social well-being, economic growth, maintenance of natural resources, and respect for the environment. To guarantee all this, it is necessary to fulfill ethical principles and moral responsibility, touching on fundamental elements for eco-sustainability such as maintaining existing resources and the planet's environmental balance. Nowadays there are environmental problems deriving from the poor way in which entrepreneurial, social, economic, and political systems have been designed, for which a notable change in the design concept is fundamental, in such a way as to allow a better coexistence with the ecological and social systems on which we depend. In the construction industry, the current challenge is to address these issues using environmentally friendly materials and construction processes, fulfilling the social and economic functions of the building in full respect of the environment.

IntechOpen

2. Wood in construction

Wood is one of the oldest building materials. In the past the great availability of material, the ease of processing and handling, and renewability, combined with specific qualities, have made wood the building material par excellence, for furniture, structural use, provisional, etc. Over time, wood technology and engineering have undergone an extraordinary evolution, but this domination has come to a halt since the nineteenth century, especially in countries where reinforced concrete and steel have monopolized the market. The reasons for this downgrading of wood are due to various cultural, economic, and environmental factors. These include natural degradability, combustibility, shape and size restrictions, and general distrust of a natural material that has inherent defects that affect its mechanical performance. Wood compared to other construction materials is characterized by the fact that the environmental conditions of installation can favor degradative or destructive alterations of biological origin to which the other materials are not prone, since they can instead be deteriorated by strong changes in temperature or by the action of gases and different chemical products that, on the contrary, leave the wood almost unaltered. The environmental hygrometric conditions do not have a significant influence on metallic and stone materials, while for wood they interfere with the values of mechanical strength. Furthermore, wood is characterized by a high susceptibility to fungus and insect attacks. Added to this is its combustibility. All these features have led to a negative view of wood as a natural construction material resulting in a setback in its use and its downgrading.

In recent decades, fortunately, there has been an extraordinary turnaround of this material in structural use even in countries where its use had been downgraded or abandoned. A great stimulus is derived from the production of laminated timber, an industrial product that incorporates the greatest advantages of wood as a natural material with the reliability, constancy of production, and high performance of an industrially produced material, and therefore subjected to quality and process control and standardization.

The advantages of wood in construction are various [2, 3].

Wood is a material characterized by good mechanical strength both in tension and in compression, so it can be used for the manufacturing of elements prone to bending such as beams, compressed like the pillars, stretched like tie rods, without the need to combine it with other materials, unlike for example concrete and masonry.

Wood is a material with a high strength to weight ratio, of the same order as steel, and it offers a compressive strength of the same order as reinforced concrete. Lightness is an important feature from the point of view of both strength and cost-effectiveness. In terms of strength, the reduced mass put into play by timber structures, for example, with respect to masonry, makes them less affected by seismic actions. In terms of cost-effectiveness, the use of a light material guarantees ease of manufacturing and handling and transport and minimization of the cost of supporting or foundation structures. Furthermore, the possibility of dry construction process allows rapid execution, modularity, and high degree of prefabrication.

Wood contributes to environmental comfort due to its low conductivity, high thermal inertia, and natural hygroscopicity. Natural materials such as wood or cork are already comfortable at room temperature, while those like stone or cement are perceived as comfortable only at higher surface temperatures.

Last but not least, the use of wood matches well the current trends of eco-compatible and sustainable construction as it minimizes the environmental impact at all levels as it is recyclable, renewable, biodegradable, and free of toxic contents.

3. Timber buildings and sustainability

The climate changes observed in recent decades, the global warming, and the increasingly frequent natural catastrophes are under everyone's eyes. In this context, the concept of climate protection summarizes all the possibilities that allow combating global warming as well as all measures to mitigate its effects.

The carbon cycle is the biogeochemical cycle through which carbon is exchanged between the geosphere, the hydrosphere, the biosphere, and the Earth's atmosphere. The current balance of the carbon cycle shows that CO_2 that enters the atmosphere is higher than the one that comes out. Much of the emissions due to human activities cannot be balanced by the natural absorption of the oceans and terrestrial ecosystems such as forests and soils, and the increase in CO_2 and other greenhouse gases leads to consequences such as the increase in global average temperature. In this contest, regarding climate protection there are basically two possible approaches: the reduction of climate-changing emissions (CO_2 and other greenhouse gases) and the expansion of carbon reservoirs, the term carbon reservoir meaning any stored form of CO_2. Wood allows both roads to be traveled [4, 5]. Thanks to photosynthesis, a tree can store large quantities of CO_2 in the wood. It is estimated that 1 m^3 of wood stores approximately 1 ton of CO_2, which remains stored in the material throughout its life, even when the wood undergoes the transformations that make it a semi-finished product or a finished product, for example, for the building industry, starting from the raw material. Therefore, the use of wood in construction involves two fundamental advantages: on the one hand, the gradual replacement of the most energy-intensive and polluting building materials with timber reduces the climate-changing emissions associated with the production and management of these materials; on the other hand, the management of forests aimed at the use of wood in industrial sectors such as buildings involves the continuous renewal of the forest itself with an increase in the capacity to extract CO_2 from the environment.

In addition to the strictly environmental sustainability aspects, the impulse to use wood in construction involves aspects of economic and social sustainability. The possibility of implementing short supply chains based on the use of local wood, for example, makes it possible to reduce the energy consumption linked to transport from distant areas of growth and at the same time allows to create employment in areas far from urban centers and therefore often economically depressed. The wood market is constantly expanding given the considerable possibilities of use, the high performance, and the technological progress of this material. The impulse to use timber in construction, especially if connected to short-term production chains, can allow:

a. The reduction of the costs of timber structures, thanks to the possibility of producing them on site instead of importing them from afar, thus promoting the sustainable building sector with great environmental advantages.

b. An increase in the demand for structural timber, with a consequent increase in forest surfaces for environmental, tourist, and hydro-geological protection advantages.

c. The creation of new jobs in the structural timber supply chain—activities related to forest care; new plants for the production of sawn wood (sawmills) and preparation of structural components (prefabrication workshops); new specialized assembly companies, with important repercussions in terms of employment; and a significant social function of reducing unemployment and abandoning depressed areas.

4. Conclusions

Wood is a natural building material that, when used in construction elements, can simultaneously fulfill the structural and esthetic function. Its use in the building sector allows it to simultaneously satisfy the aspects of environmental, economic, and social sustainability, especially when it is accompanied by implementation of short supply chain processes that allow the use of local wood and processes for controlling the sustainable management of forests that avoid deforestation and the impoverishment of the territories.

In the current context of implementing strategies to combat climate change and in general to protect the environment, the characteristics of this material and its qualities suggest giving impetus to its use in highly polluting sectors such as construction.

Acknowledgements

The author would like to acknowledge Sardegna Ricerche for the financial support (POR FESR 2014/2020—ASSE PRIORITARIO I "RICERCA SCIENTIFICA, SVILUPPO TECNOLOGICO E INNOVAZIONE").

Author details

Giovanna Concu
Department of Civil and Environmental Engineering and Architecture, University of Cagliari, Cagliari, Italy

*Address all correspondence to: gconcu@unica.it

References

[1] Brundtland GH. Our Common Future: Report of the World Commission on Environment and Development. Geneva: UN-Dokument A/42/427; 1987

[2] Asdrubali F, Ferracuti B, Lombardi L, Guattari C, Evangelisti L, Grazieschi G. A review of structural, thermo-physical, acoustical, and environmental properties of wooden materials for building applications. Building and Environment. 2017;**114**:307-332

[3] Nässén J, Hedenus F, Karlsson S, Holmberg J. Concrete vs. wood in buildings—An energy system approach. Building and Environment. 2012;**51**:361-369

[4] Hildebrandt J, Hagemann N, Thrän D. The contribution of woodbased construction materials for leveraging a low carbon building sector in Europe. Sustainable Cities and Society. 2017;**34**:405-418

[5] Ramage MH, Burridge H, Busse-Wicher M, Fereday G, Reynolds T, Shah DU, et al. The wood from the trees: The use of timber in construction. Renewable and Sustainable Energy Reviews. 2017;**68**:333-359

Chapter 2

Lumber-Based Mass Timber Products in Construction

Meng Gong

Abstract

This chapter provides information related to commonly used wood construction methods (i.e., light-frame, post-and-beam, and mass timber) and mass timber products. It briefly discusses the manufacturing of four major lumber-based mass timber products (i.e., glue-laminated timber, nail-laminated timber, dowel-laminated timber, and cross-laminated timber), and their available dimensions and typical applications. The discussion also addresses primary lumber products, such as dimension lumber, machine stress-rated lumber, and finger-joined lumber, which are the building blocks from which mass timber products are manufactured. Advantages of using wood in construction are illustrated by examples largely from North American practices. The life cycle assessment concept is also introduced.

Keywords: building materials, cross-laminated timber, dowel-laminated timber, glue-laminated timber, life cycle assessment, lumber, mass timber products, nail-laminated timber, timber, wood construction

1. Introduction

Prior to the availability of rolled steel and reinforced concrete, wood was the primary structural material in North America and other timber-rich regions of the world [1]. However, the raw material resources keep changing in more recent times, e.g. log diameters become smaller and trees come from faster growing plantation species. As a result, traditional solid timber products have been supplemented by Engineered Wood Products (EWP) like glue-laminated timber (GLT or glulam), laminated veneer lumber (LVL), and oriented strand board (OSB). This has permitted economic construction of residential and nonresidential buildings, bridges, and industrial structures. Presently, using of traditional wood products and EWP is recognized as a "green" option, and is encouraged by governments as part of sustainable development and climate change mitigation strategies [2].

From a technical perspective, modern EWP commonly provide better and more predictable physical and mechanical properties than traditional wood products, such as more uniform structure, greater dimensional stability, greater strength, and stiffness. Initially, much development of EWP was focused on creating substitute products capable of replacing small dimension sawn lumber and boards as primary elements in light-frame building superstructures; but in recent decades, much attention has been switched to creation of mass timber products (MTP). The term MTP describes a family of EWP of large section size that offers the construction industry a viable alternative to use structural steel and reinforced concrete [3]. This includes thick-panel products, such as cross-laminated timber (CLT) and structural composite lumber (SCL), as well as adhesively or mechanically laminated linear elements like GLT, nail-laminated timber (NLT), and dowel-laminated timber (DLT).

SCL refers to products manufactured by layering dried and graded wood veneers or strands bonded together by moisture-resistant adhesive into panel-like products of a width of up to 2.44 m, a thickness of 38 mm, or more. In principle, SCL is only limited in width and length by transportation considerations. SCL basically includes LVL, laminated strand lumber (LSL) and oriented strand lumber (OSL), which is usually sawn into lumber-like products. However, parallel stand lumber (PSL) is also deemed as a SCL product, which is commonly used as columns connected to other MTP. Use of terms in the literature can be colloquial, with timber-concrete composite (TCC) and other hybrid elements sometimes grouped into the meaning of MTP. Overall, MTP offers architects and builders many opportunities to express their concepts creatively, while satisfying various technical performance requirements applicable to engineered structures of many types [4]. This chapter places emphasis on the types of lumber-based MTP illustrated in **Figure 1**.

Figure 2 illustrates three types of wood construction methods, namely light-frame, post-and-beam, and mass timber. Light-frame construction consists of studs, joists, and other framing at spacings of 600 mm or less [6], **Figure 2**—top. Dimension lumber is used for framing members and plywood or OSB for sheathing materials. Light-frame construction is an economical choice for the construction of low- and mid-rise buildings, which makes use of dimension lumber in a range of grades and dimensions [7]. Light-frame wood structures can be also used for shopping centers, plazas, service and maintenance buildings, and institutional and municipal facilities. Prefabrication of components such as wall and roof panels, even complete homes or office units are efficient extensions of this framing technique [7]. However, the structural system of a light-frame building is not well-defined, resulting in much redundancy. Design of a light-frame building often only includes architects unless the building is large.

Post-and-beam construction is a skeletal framework of posts, beams, and decking supported on a foundation, in which the posts and beams are well spaced apart, more

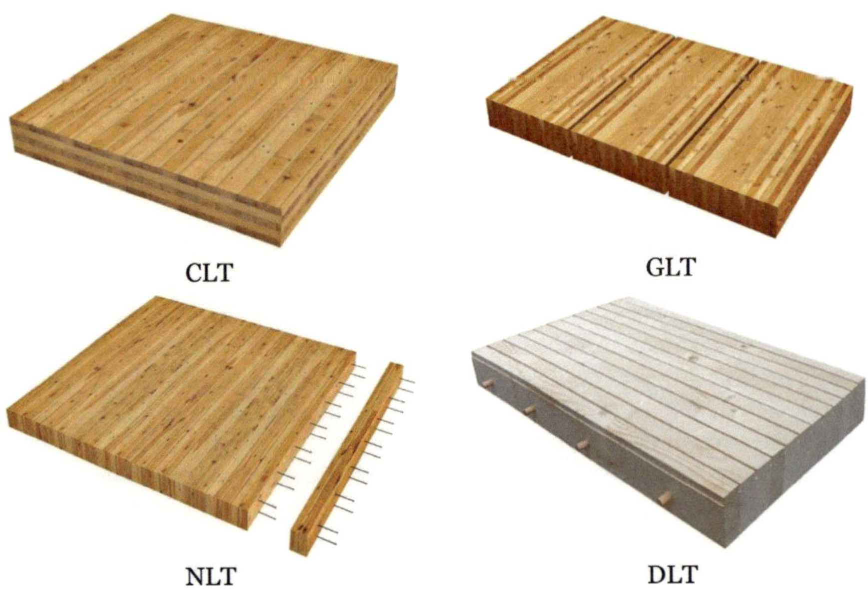

Figure 1.
Lumber-based MTP (Source: Images obtained from StructureCraft).

Figure 2.
Wood building construction methods. Top: Light-frame construction (Source: Photo obtained from Okoye et al. [5]). Center: Post-and-beam construction (Source: Photo obtained from Post & Beam Homes). Bottom: Mass timber construction (Source: Photo obtained from UBC Public Affairs).

than 600 mm, but commonly 1200 mm or more [6], **Figure 2**—center. Traditionally, posts and beams were made of large solid timbers, which were connected with mortise and tenon joints locked into place with hardwood pegs, with diagonal braces for stabilization of a structure [7]. Nowadays, many types of EWPs (such as GLT and LVL) and connectors (such as metal brackets, shear plates, and split rings with bolts) are very often used. The post-and-beam construction is commonly used to construct custom-designed homes, commercial buildings, recreation centers, and industrial structures, for reasons of ease of fabrication and consequent economy [7]. Unlike the light-frame construction, the structural system of a post-and-beam building is well-defined and engineered, generating very limited or no redundancy. Design of a post-and-beam building is usually formal, involving both architects and engineers, especially if the building is relatively large. Hybrid post-and-beam and light-frame construction features the exposed heavy timber components, but allows insulation to be placed in the wall space, with finishes applied to both the inner and outer faces of the studs [7].

Mass timber construction complements traditional light-frame and post-and-beam construction methods due to emergence of various types of MTPs, **Figure 2**—bottom. It creates single or multiple material hybrid superstructures for building and other structures. Since beams are not always required, new technology and terminology, such as post-and-panel construction, have emerged. This demonstrates that MTPs have been developed into material options, where the only limits on their uses are limitations of the inventiveness of minds of architects and engineers within the scope of what applicable building/construction regulations permit [4]. What codes and standards permit architects and engineers to do is not yet uniform; but in the broad sense, construction codes and standards in various countries have transitioned, or are transitioning, away from prescriptive provisions to performance-based provisions in a manner that enables greater use of EWP, including MTP. Most important in this respect is the revision of fire performance provisions related to buildings [4, 6]. Mass timber systems are widely reported to be cost-competitive, carbon-efficient, sustainable and reliable, which stem from the scientific data generated from full-scale fire, seismic, durability, acoustic, and vibration tests being conducted internationally by researchers and engineers [3, 4]. It is now reasonable to claim that the use of EWP and MTP has the same level of supporting technical understanding as that underpinning any other major class of construction material. Latter sections of this chapter demonstrate the use of MTP as parts of high-performance buildings meeting needs of society and occupants.

2. Lumber and lumber-based MTP

2.1 Lumber

Lumber is a manufactured product derived from logs, including boards (elements with limited thickness), dimension lumber (elements with relatively small section dimensions), and timbers (elements with relatively large section dimensions). In North America, most lumber is softwood dimension lumber having thicknesses ranging from 38 to 89 mm, widths from 38 to 184 mm, and lengths of up to about 5 m [6]. Dimension lumber is widely used in light-frame construction, which is categorized into four groups in the Canadian practice: structural light framing, structural joists and planks, light framing, and studs. Dimension lumber is usually graded by visual inspection in terms of appearance characteristics, such as knots and slope of grain. For example, the grades of dimension lumber used for structural light framing construction are Select Structural (SS), No. 1, No. 2, and No. 3. It should be noted that there is not a strength difference between No. 1 and No. 2 Canadian dimension lumber albeit there exists an appearance difference [6]. Therefore, the product mix of No. 2 and Better is commonly used where the appearance of No. 1 grade lumber is not required. Alternatively, dimension lumber can be mechanically evaluated and sorted into grades using so-called machine stress-rated (MSR) lumber or machine-evaluated lumber (MEL) [6]. The MSR machine is widely used in wood industry to nondestructively test each piece of dimension lumber to determine its stiffness so that it can be assigned a permitted design stress based on the established relationship between the stiffness and bending strength. In North America, grades of MSR lumber are assigned "f-E" values, such as 1950f-1.7E. The "f" value designates the predicted strength in pounds per square inch (psi), and the "E" value designates the average stiffness measured in millions of pounds per square inch (10^6 psi) [6].

Boards are lumber products having thicknesses of 32 mm or less, making them usable as decking and sheathing. When the smallest cross-sectional dimension of a lumber product reaches or exceeds 140 mm it is termed timber, which is graded based on visual inspection methods [6]. Uses of dimension lumber and timbers widely range with differences in whether the former or latter is suitable depending on the type of structural system, and performance requirements applicable to a structural system. In general, dimension lumber is used in systems where multiple parallelly arranged elements act together to resist effects of particular structural design loads. Timbers, on the other hand, can be used in situations where multiple elements or a single element is designed to resist effects of particular structural design loads. Another important difference is that dimension lumber elements must always be protected from effects of design fire situations; whereas, depending on specifics of a situation, timbers may not require such protection.

Finger-joints are commonly used to join short pieces of lumber together to make longer pieces. Meshing wedges known as "fingers" are made as either side of a joint, as illustrated in **Figure 3**, and bonded using structural adhesive. The joint profile governs the strength of a joint, and is defined by the finger length, tip thickness, tip gap, and finger pitch, slope, and depth. For example, a 29-mm-long finger joint is commonly used (**Figure 3**—left). However, reducing finger length to 13 mm with some modifications to the joint profile (**Figure 3**—right) not only helps to reduce material waste, but also keeps the same or slightly higher strength joints [8]. Also, it is noted that cutting out strength reducing features like large knots then finger joining lumber is a highly effective way of upgrading properties of dimension lumber, increasing value, and enabling higher value uses like creation of high-performance MTP [6]. Another advantage of finger joining lumber is that it increases dimensional stability under changing environmental conditions prior to or after installation of lumber in structures. Adhesives used in finger-joints are usually phenol-resorcinol formaldehyde for lumber products intended for general applications or incorporated in GLT elements, or polyvinyl acetate for lumber products used as studs [6].

2.2 Glue-laminated timber (GLT or glulam)

GLT (also widely known as glulam) is a structural product composed of multiple pieces of finger-joined dimension lumber, or other types of EWP, adhesively face-to-face bonded to create a desired form. GLT was first used in Europe in the early 1890s. A 1901 patent from Switzerland signaled the true beginning of GLT construction [9]. A significant development in the GLT industry was the introduction of fully

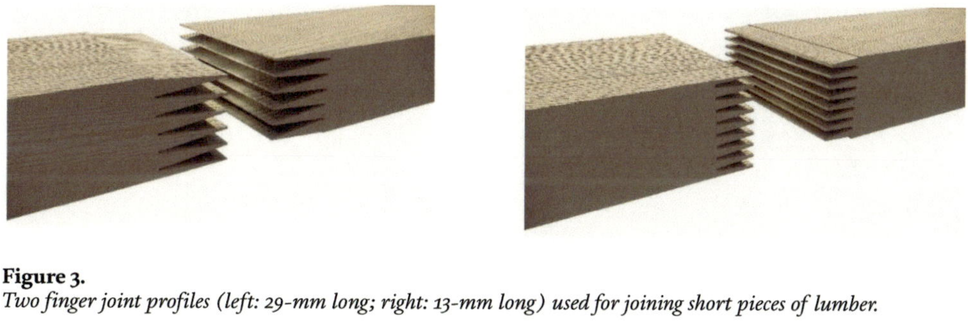

Figure 3.
Two finger joint profiles (left: 29-mm long; right: 13-mm long) used for joining short pieces of lumber.

water-resistant phenol-resorcinol adhesives in 1942, which allowed GLT to be used in exposed exterior environments without concern of glueline degradation [9]. The manufacturing of GLT is deemed as a one-dimension additive process. The grain of all laminations runs parallel with the lengths of straight members, **Figure 4**. The dimension lumber laminations are not visually graded on the same rules as regular lumber, but follow the grading rules stipulated in Canadian Standard O122 "Structural Glued-Laminated Timber" [7]. Each lamination is visually inspected based on both faces of the piece, and then assigned one of four grades: B-F, B, D, or C [7], in which B-F indicates the highest grade and C the lowest grade. Laminations of higher grades are used in the top and bottom portions of a GLT beam, **Figure 5**, where bending stress is greatest. Specified laminations are also nondestructively graded by machine before assembly to meet both visual and stiffness requirements for particular grades of GLT. Sometimes layers of other materials, such as glass fibers, are incorporated among lumber laminations to add strength or stiffness or to locally reinforce GLT [10]. Moisture content of laminations ranges from 7 to 15% during fabrication. Durable cold-setting waterproof-structural adhesives are used, such as phenol formaldehyde and phenol-resorcinol formaldehyde [7]. Because finger-joined lumber is employed, dimensions of GLT members are in principle only limited by manufacturing and transportation capabilities of a manufacturer. Those capabilities are highly variable, with the most advanced involving fully automated manufacturing processes based on advanced integrated design and manufacturing methods. The automated processes can include robot handling of materials and elements from the arrival of logs at a manufacturing plant to installation of elements at a construction site. A typical GLT member ranges in depth from 114 to 2128 mm or more, in width from 80 to 365 mm, and in length of up to 40 m [7]. GLT is commonly used as beams and columns (**Figure 4**—left and middle), but can be also used as flexural members (**Figure 4**—right). In latter situation, the narrow faces of the laminations are normal to the direction of the load. The Canadian Standard O86 "Engineering Design in Wood" refers to this condition as "vertically glue-laminated" [11]. Usually, GLT is used in dry service conditions or is protected in some way if used under outdoor conditions.

Design stiffness and strength properties of GLT of a given grade are calculated based on engineering properties of the laminations using equivalent linear elastic mechanics theories. A wide range of GLT grades are available with some involving deliberate placements of laminations of different grades to achieve the design properties of GLT elements suited to their particular applications [11]. In general, there are two grade categories for GLT, stress grade and appearance grade [7]. The former defines specified strengths of a GLT member, and the latter the quality of

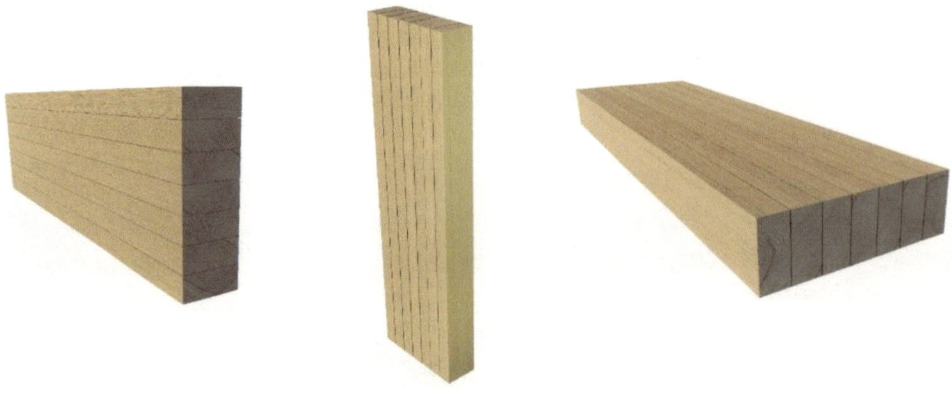

Figure 4.
GLT beam (left), column (middle), and panel (right).

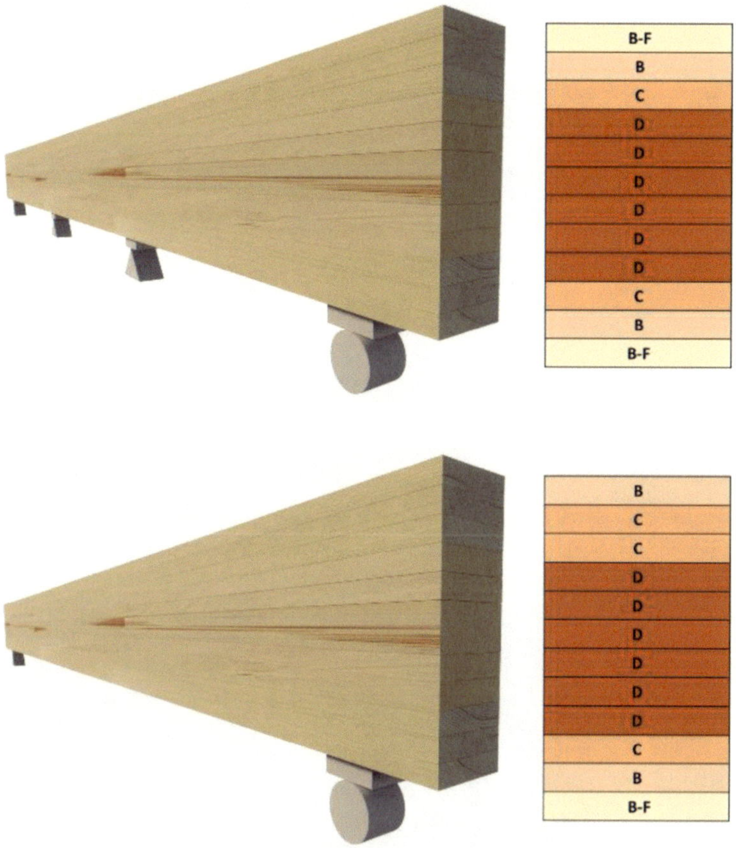

Figure 5.
GLT members with laminations suiting resistance of bending forces: Upper—Member with a balanced layup intended to maximize material use when the member is loaded in tension on both top and bottom faces (EX grade under Canadian system), and Lower—Member with an unbalanced layup intended to maximize material use when the member is loaded in tension on the bottom face (E grade under Canadian system).

finish on the exposed surfaces of the member. For example, some grades suit uses of GLT elements as beams, columns or tension members, **Figures 6–8**. Taking the Canadian Standard O86 "Engineering Design in Wood" as an example, that design standard specifies the grades of GLT bending elements as 20f-E, 20f-EX, 24f-E, and 24f-EX [11], **Figure 5**. Within those designations, numbers 20 and 24 are indicative of the associated specified design strength in bending. E indicates that associated grade properties apply to elements without an inflection in their deformed shapes, with the proviso faces intended to be stressed in tension are correctly oriented. EX indicates that associated grade properties apply to elements with inflections in their deformed shapes. Similar approaches are adopted by other international standards which define rules for engineering design of timber structures.

2.3 Nail-laminated timber

NLT is manufactured with dimension lumber laminations, stacked on edges, and fastened with nails, to create large-flat structural components, **Figure 9**. Spikes and screws are sometimes used as well. Since the beginning of the nineteenth century, NLT systems were utilized as floor elements in structures known as "mill construction" that originated from cotton mills and sawmills found in the North Eastern United States [7]. The prevalence of the industrial building systems led

Figure 6.
Conference room built with GLT beams and columns at the University of New Brunswick, Fredericton, Canada.

the National Lumber Manufacturers Association to publish a guide "Heavy Timber Mill Construction Buildings" in 1916 [7]. In addition, NLT has been used to create deck and diaphragm elements of bridges and buildings for centuries [6]. Like GLT, the manufacturing of NLT is a one-dimension additive process. In North America, individual laminations have a thickness of 38 mm or more and a depth of 64 mm or more, similar to plank decking [7]. The moisture content of laminations is usually 12–16% at time of manufacturing of NLT [14]. The visual or MSR grade of softwood laminations are widely used, such as SS and No. 2 and Better or 1650f-1.5E [7, 14]. Single laminations are commonly employed if the length of prefabricated panels is less than 6 m [14]. The spliced laminations of specific pattern [11, 14] or finger joined lumber laminations [14] are used if longer panels are required. The Canadian Standard O86 "Engineering Design in Wood" [11], for example, specifies connection requirements for fabricating NLT, requiring that nails be long enough to pass through two adjacent laminations and at least halfway through the third, **Figure 9**. For example, 102-mm-long nails should be used to fasten 38-mm-thick laminations, and 152-mm-long nails for 64-mm-thick laminations. Such requirements are based on practical experience and ensure integrity of NLT in various end use situations. NLT shall be spiked together with a staggered single row of nails at intervals of not

Figure 7.
A forest of intertwined GLT trees in the Carlo Fidani Peel Regional Cancer Centre, Mississauga, Canada (Source: Photos obtained from CWC [12]).

more than 450 mm [7]. The prefabricated NLT panels typically come in lengths of 3–8 m; however, the panel size is limited by transportation restriction [7]. The drawbacks of using NLT are its slow fabrication process and after-fabrication machining problem due to existence of nails.

In North America, many timber decks of rural bridges constructed from 1920s through the mid-1960s were made of NLT [15]. Mostly, the NLT was oriented so the lumber laminations were transverse to the bridge span and supported by bridge girders, but for short bridges lumber laminations were sometimes orientated parallel to the span [15], **Figure 10**. Another common traditional use of NLT is in floors of industrial and commercial buildings. The reasons for choosing NLT are as follows: it is well suited to onsite fabrication; it is capable because of the nails of absorbing energy damping vibrations caused by transient or sustained dynamic force (e.g., bridge wheel loads and reciprocating industrial equipment); and it has good fire performance. Disadvantages of NLT include that it is not particularly mechanically efficient if NLT elements are required to have high rigidity when loaded in-plane or as flexural elements, also there have been durability issues associated in particular with bridge applications. The disadvantages stem from the flexibility of nailed interconnections between laminations, and proneness to gaps to form at those interconnections (e.g., due to moisture movements in the laminations).

Figure 8.
160-m-long timber bridge with GLT deck structure in Mistissini, Canada. (Source: Photos obtained from Lefebvre and Richard [13]).

Figure 9.
NLT with linear (left) and staggered (right) nailing patterns.

Figure 10.
Transverse (left) and longitudinal (right) oriented NLT bridge decks.

Recently, use of NLT has undergone resurgence as part of the modern mass timber movement in buildings [3, 4], **Figure 11**. This, in some cases, supports adoption of complicated architectural forms, **Figure 12**, supported by creation of hybrid NLT products which combine lumber laminations with layers of sheathing materials such as plywood and OSB to reinforce the system [14, 17]. Sheathing adequately nailed to NLT can create a diaphragm of the capability to resist lateral forces, and can also help keep the system dry if exposed to moisture [7]. In any such case, it is required to consider the system as an individually designed engineering project.

Figure 11.
Seven-story T3 Minneapolis building made of NLT floors and GLT beams and columns in USA (Source: Photos obtained from StructureCraft [16]).

Figure 12.
Qingdao Pearl Visitor Centre of exposed NLT roof in China (Source: Photo obtained from StructureCraft [16]).

2.4 Dowel-laminated timber

Dowel-laminated timber (DLT) is another member of MTP family. DLT is similar to NLT regarding laminations, but different in fasteners. Instead of nails, DLT uses hardwood dowels to join laminations, **Figure 13**. The manufacturing of DLT is another example of one-dimension additive process. DLT was developed in the early 1990s in Switzerland [7]. DLT is manufactured with softwood lumber of a thickness of 38 mm and a depth of 89, 140, or 184 mm, stacked on edges just like NLT, and fastened face-to-face with wooden dowels. Unlike NLT, finger-joined lumber is typically used in manufacturing of DLT. The moisture content of laminations is 19% or less at time of manufacturing [7]. The visual or MSR grade of laminations is, if spruce-pine-fir lumber is used for example, SS and No. 2 and Better or 2100f-1.8E [18]. The wooden dowels, which are usually made of high-density hardwood species (such as oak), have typically a diameter of 19 mm and a moisture content of approximately 6–8% [7]. The predrilled holes of a diameter being about the same as dowels are required prior to driving dowels into laminations [18]. The dowels can then be hydraulically pressed in a linear or staggered way with spacing of 300 mm [7], the latter of which could offer additional stiffness DLT panels, **Figure 13**. Dowels

Figure 13.
DLT with linear (left) and staggered (right) fastening patterns.

are commonly penetrated through 7–10 laminations, resulting in a more efficient process of manufacturing DLT than NLT. As the moisture content of both materials used in DLT equilibrate after fabrication, the dowels swell and the lumber shrinks, which forms a strong friction-fit joint between the lumber and the dowels, resulting in a panel that does not require glue or nails [7]. DLT has been gaining interest in both Europe and North America since it is almost made of 100% wood, except those of finger-joined laminations that contains very limited amount adhesive. DLT is ease of being manufactured using computer numerical controlled (CNC) machinery, such as lathes, routers, and mills. The prefabricated DLT panels typically have a length up to 18 m, a width up to 4.3 m in any increment, and a thickness ranging from 76 to 349 mm [18]; however, the panel size is usually limited by transportation restrictions.

DLT panels can readily be milled and routed for preintegrated electrical and other service conduits, which offers a unique feature to DLT, i.e., the flexible design, **Figure 14**. This also allows designers to improve the acoustic performance and visual appeals of a building by making kerfs and curves. For example, acoustical strips can be integrated into the bottom surface of a DLT panel, helping designers reduce sound while keeping the wood exposed and also allowing for a variety of surface finishes [18]. DLT can be also recognized as a type of MTP that can be used in exterior exposure, allowing itself to be used for decks, balconies, and canopies.

DLT performs similarly, in terms of structural performance, to GLT and NLT, because its grains run in one direction. DLT allows a significant flexibility in architectural design, which is well suited for floor and roof applications, but can be used as wall panels as well [18], **Figure 15**. Two-way spans can be achieved with the use of reinforcement such as adding multiple layers of plywood atop the DLT panels [18]. In addition, DLT panels can be used as structural bearing or shear walls, and elevator and stair shafts. The design requirements for DLT may be considered the same as those used for NLT, if the hardwood dowels can adequately connect the laminations [7]. In reality, there is almost nothing that can be referenced in the

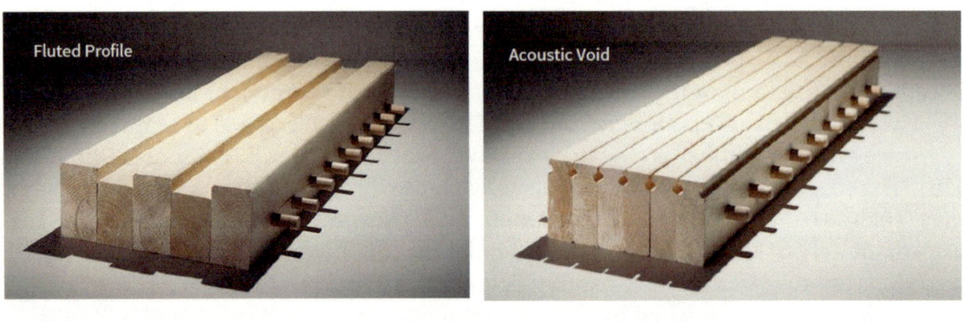

Figure 14.
Two sample profiles of DLT (Source: Pictures obtained from StructureCraft [18]).

Figure 15.
Seven-story T3 Atlanta building comprising DLT floor and roof panels in USA (Source: Photos obtained from StructureCraft [18]).

codes worldwide, except that a few manufacturers provide published design values for their own DLT products [4]. Thus, use of DLT would require approval by the building authority on a case-by-case basis.

2.5 Cross-laminated timber

Cross-laminated timber (CLT) is a new-generation engineered large-size structural panel product, which consists of layers of dimension/MSR lumber (typically three, five, or seven) oriented at right angles to one another and then bonded using adhesives, **Figure 16**—upper. CLT was originally invented in the 1970s in Europe [6] and introduced as an innovative wood product in the early 1990s in Austria and Germany [19]. In the mid-1990s, Austria undertook an industry-academia joint research effort that resulted in the development of modern CLT [19]. In the last 2 decades, the use of CLT has gained interest to both construction and wood industries in North America, featured with the publication of two editions of CLT Handbook [20, 21] and erection of 18-stories CLT building "Brock Commons Tallwood House" in 2017 in Canada. Unlike GLT, NLT, and DLT, the manufacturing of CLT is a kind of three-dimension additive process. The species of wood used depends on the location of a manufacturing plant. For example, black spruce is widely used in Eastern Canada. The commonly used lumber products in manufacturing of CLT are dimension lumber of a grade of No. 1/No. 2 or MSR lumber of a grade of 1200f-1.2E or better in its major strength direction, and dimension lumber of a minimum grade of No. 3 in its minor strength direction [7]. In the major strength direction, the minimum net width of a lamination shall be 1.75 times its thickness, and in the minor strength direction, the net width of a lamination shall not be less than 3.5 times its thickness if the laminations are not edge-glued [7]. The moisture content of lumber at fabrication of CLT is about 12% [7]. The cold-set structural adhesives are preferred to increase the productivity of manufacturing CLT panels, which include emulsion polymer isocyanate (EPI), polyurethane (PUR), and phenol-resorcinol formaldehyde (PRF) [7]. In Canada, the adhesives used in manufacturing process of CLT must comply with the Canadian Standard O112.10 "Evaluation of Adhesives for Structural Wood Products (Limited Moisture Exposure)" and ASTM D7247 "Standard Test Method for Evaluating the Shear Strength of Adhesive Bonds in Laminated Wood Products at Elevated Temperatures" [7]. The finger- or scarf-joined lumber is used to face-to-face and/or edge-by-edge laminating as two-dimension components. Use of edge-gluing or not slightly differs in the manufacturing of CLT between North America and Europe. In North America, edge-gluing of lumber is not a common practice due

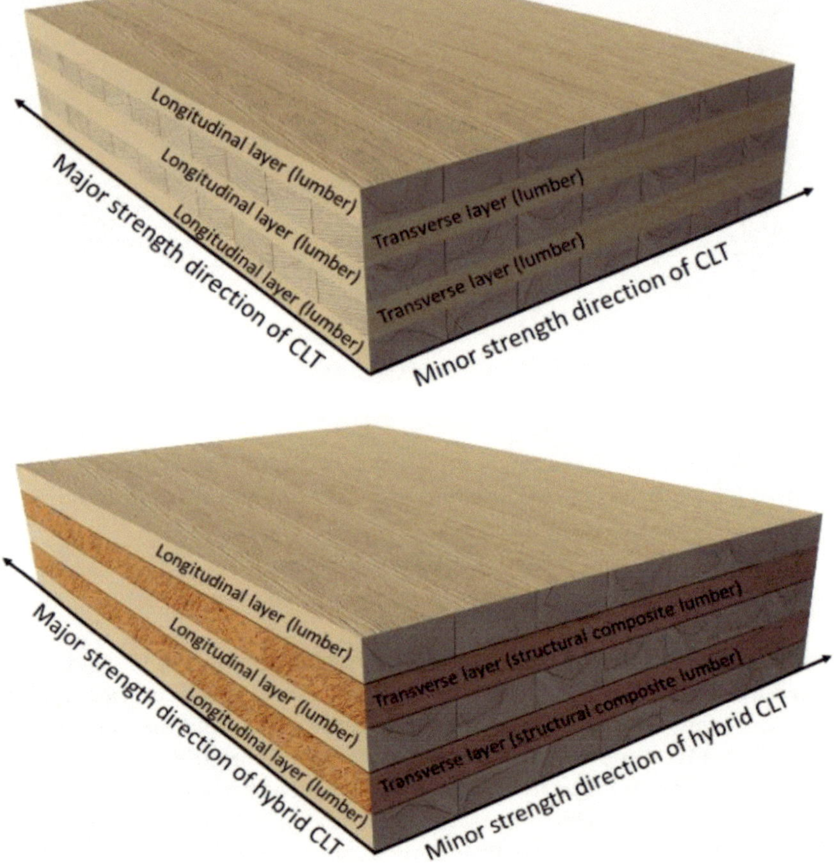

Figure 16.
CLT products (Upper: a generic CLT made of lumber only; lower: a hybrid CLT made of dimension lumber in the major strength direction and structural composite lumber in the minor strength direction).

to the added manufacturing costs. The gaps between lumber could provide some tolerances for wood movement due to the change in moisture in service. However, the European practice appears to widely apply edge-gluing with an aim to offer good stiffness and strength of a CLT panel. Anyhow, as a trade-off between cost and improved panel performance, edge-gluing of selected layers as needed could be adopted [6]. CNC routers are often employed to precisely cut CLT panels to size and openings for windows, doors, connections, ducts, and service channels. A CLT product can be produced in large sizes of a width ranging from 1.2 to 3 m, a length from 5 to 19.5 m, and a thickness from 100 to 500 mm [7]. CLT can be also manufactured in custom dimensions, with panel sizes varying by a manufacturer.

Despite the availability of commercial machines to manufacture construction size CLT using dimension lumber, there are challenges with the existing systems, such as the need to apply pressure to all four sides of a panel to ensure adequate edge-glue bond quality, as well as the out-of-plane pressing. From a product performance perspective, CLT is known to be prone to the so-called rolling shear failure and excessive deflection when subjected to out-of-plane loading. This is particularly critical where the lumber layers are not edge-glued. These performance issues could be addressed by replacing one or more of the layers in a CLT panel with SCL, such as LSL and OSL. Such an innovative hybrid CLT can offer many advantages over the generic one that is made of 100% dimension lumber, **Figure 16**—lower. The hybrid CLT products could reduce the production cost because of the reduced efforts to layup of individual lumber pieces and the possible elimination of the need to press the panel on all four sides simultaneously, improve the rolling shear strength and stiffness properties of generic CLT since SCL has relatively high shear strength and rigidity, and improve the fire resistance of CLT due to the elimination of gaps present in generic CLT made with non-edge-glued dimension lumber. The research on three- and five-layer hybrid CLT, recently conducted in the Wood Science and Technology Centre, the University of New Brunswick, Canada, showed that the bending stiffness, moment capacity, and shear capacity of hybrid CLT were increased to a large degree in comparison to generic one [22–24].

Cross laminating technology provides CLT panels with improved stable dimensions, and relatively high in-plane and out-of-plane stiffness and strength properties in both directions, giving these panels a two-way action capability [6]. It is well-suited to floors, walls, and roofs, and may be left exposed on the interior for esthetics. The light weight of CLT directly helps reduce the size and cost of foundation. As a prefabricated building component, CLT offers shorter onsite construction time than traditional platform frame construction or steel and concrete construction, minimizes waste and noise during construction, and provides a very competitive cost in comparison to concrete and steel [19]. CLT has also been used to fabricate bridge decks, heavy equipment mats, and platforms for oil rigs, and to construct mid-rise and tall wood buildings of over seven stories, and large industrial structures [4]. In addition, CLT exhibits good seismic and fire performance. The 2015 International Building Code (IBC) and 2015 International Residential Code recognize CLT products manufactured according to the ANSI/APA PRG-320 "Standard for Performance Rated Cross-Laminated Timber." Under the 2015 IBC, CLT at the required size is specifically stated for prescribed use in Type IV buildings, i.e., heavy timber buildings, which hold well under fire conditions due to formation of char layer. However, CLT can be used in all types of combustible construction, i.e., wherever combustible framing or heavy timber materials are allowed [4].

CLT is sometimes deemed as a standalone building material and construction system. A kind of post-and-panel construction has emerged, accompanied with many innovative connections. The tallest wood building as of the year of 2018,

Figure 17.
Brock Commons Tallwood House, Vancouver, Canada (Source: Photo obtained from UBC Public Affairs).

Brock Commons Tallwood House (**Figure 17**), stands in Vancouver, Canada. This building includes 17 stories of CLT floors supported on GLT columns atop a concrete base with two 18-stroy concrete cores. This 53-m-high building is used as student residence providing 404 bed units. Its unique designed column-to-column metal connector makes a column-panel-column connection, minimizing the accumulation of deformations (i.e., the transverse wood movement) generated from each CLT floor. It was reported that 80% of the work for this tall building was prefabricated and 70% alone was gaining code approval [25].

3. Endnotes

Environmental awareness coupled with sustainable design and construction practices are increasingly becoming a requirement for many building projects throughout North America and around the world [7]. Sustainable design aspires to use less energy and material resources in conjunction with lowering the environmental impacts on a building from its cradle to grave [7]. The reasons for using wood in construction are attributed to its environmentally friendly attributes, ease of assembly, reduced noise and waste during construction, natural beauty, and cost-effectiveness. Increasing use of renewable and sustainable building materials in construction, such as wood, is a worldwide move. Wood-based materials, such as MTP, consume less energy and emit fewer greenhouse gasses (GHG) and pollutants over their life cycle than traditional energy-intensive construction materials such as steel and concrete [2]. To spur innovation and certify the performance of wood as a construction material, many countries have made a great effort to support the research and development of wood products such as MTP. In Canada, for example, the 2015 Edition of its National Building Code of Canada (NBC) allows to construct wood frame buildings up to six stories. The Canadian have been working hard to the code revisions with an aim at the 2020 Edition of the NBC to permit tall wood buildings up to 12 stories [2]. Their long-term objective is to establish the performance-based codes for the 2025 Edition of the NBC and beyond, which will eliminate the distinction between building materials. This will give architects and developers freedom of choice in their materials. Ramage et al. illustrated the selection of structural systems for multi-story buildings in terms of the number of stories and their use of wood [26], **Figure 18**. For buildings up to about six stories, CLT uses substantially more wood to achieve the same function as a light-wood frame building. For buildings over six stories, the use of CLT together with light-wood frame may use less wood than CLT alone. As for buildings taller than 10 stories, the mass timber construction method is employed by using GLT megaframe to support CLT walls, floors, and roofs [26].

The life cycle of a product is defined in the standard ISO 14040 as "consecutive and interlinked stages of a product system, from raw material acquisition or generation from natural resources to final disposal" [27]. This has led to the use of the life cycle assessment (LCA), which is defined as "the compilation and evaluation of the inputs and outputs and the potential environmental impacts of a product system throughout its life cycle" [27]. LCA is a performance-based approach to assessing impacts that building products or systems have on the environment over their lifetime [7], including all activities from raw material extraction/harvesting, materials processing/products manufacturing, transportation, distribution, installation, use, repair and maintenance, and final disposal or recycling [7]. LCA is deemed as the best available tool to compare sustainability of building materials, which includes four main phases, i.e., goal and scope definition, inventory analysis, impact assessment, and interpretation.

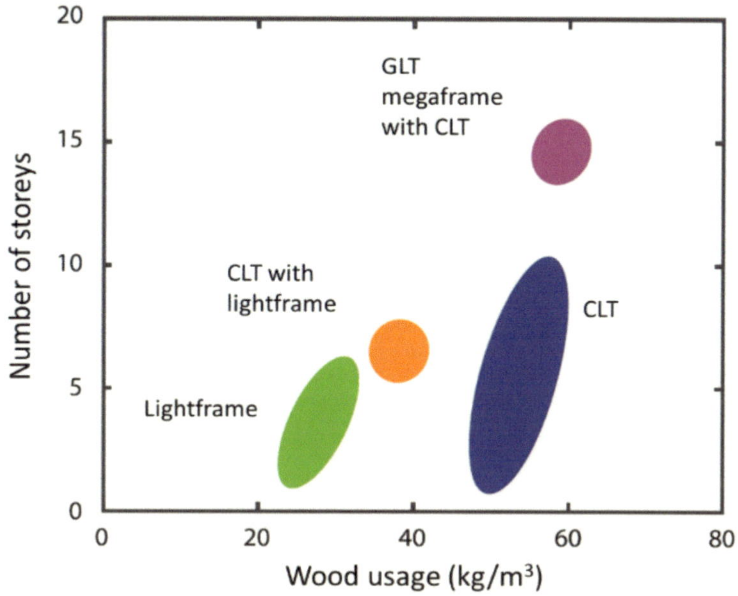

Figure 18.
Use of structural lumber and mass timber products for various structural systems (Adapted from Ramage et al. [26]).

LCA studies on wood buildings are rooted in the assumption of the same life span for wood as other structural materials. Ramage et al. summarized, after conducting a comprehensive review on use of wood in construction, that the buildings are really demolished due to degradation of their main structure, whatever the structural materials [26]. However, some wood components in a building may have a design life shorter than that of the building as a whole, or may require maintenance during the life of the building. There are many factors impacting the lifespan of wood components, including fire and natural degradation. In comparison to other building materials such as steel and concrete, wood is combustible. However, large cross-section wood components, such as those made of GLT and CLT, may perform well in case of catching a fire due to the formation of char layer that can act to insulate the material inside. The burnt wood can still keep large, enough strength to support the integrity of a building. As for small cross-section of wood components, they must be encapsulated in noncombustible material such as gypsum boards or concrete. Steel connectors are widely used in modern wood buildings, thus heat can be quickly conducted through the connectors, degrading the strength and stiffness of the wood connections and materials around them [26]. Caution must be used at time of using steel connectors in construction of wood buildings.

In a summary, mass timber building systems make it feasible to use wood in construction of mid-rise and tall buildings, industrial structures, and bridges. However, mass timber products and building systems behave in a fundamentally different way in fire than steel or concrete buildings in structural and spatial layout. More research is required to increase use of wood in construction.

Acknowledgements

This piece of work was financially supported by the New Brunswick Innovation Research Chair Program, New Brunswick Innovation Foundation, Canada.

The author's sincere gratitude goes to Dr. Ian Smith, Emeritus Professor of the University of New Brunswick (UNB), for his kindly reviewing part of the manuscript. The author's thanks also go to Mr. Luji Xiong, Graduate Research Assistant at UNB, for drawing sketches.

Conflict of interest

I confirm there are no conflicts of interest.

Author details

Meng Gong
University of New Brunswick, Fredericton, Canada

*Address all correspondence to: meng.gong@unb.ca

References

[1] Forest Products Laboratory (FPL). Wood Handbook—Wood as an Engineering Material. General Technical Report FPL-GTR-190. Madison, WI, USA: U.S. Department of Agriculture, Forest Service, Forest Products Laboratory; 2010. 509 p

[2] Natural Resources Canada (NRCan). 2019. Available from: https://www.nrcan.gc.ca/forests/industry/products-applications/16834/ [Accessed: February 08, 2019]

[3] Harte AM. Mass timber—The emergence of a modern construction material. Journal of Structural Integrity and Maintenance. 2017;**2**(3):121-132

[4] Thinkwood. Mass Timber in North America. 2018. Available from: https://www.thinkwood.com/our-ceus/mass-timber-north-america [Accessed: December 18, 2018]

[5] Okoye U, Kam-Biron M, Perkins B, Barnes C. Higher education that includes timber engineering. STRUCTURE magazine. 2017;**July**:49-51

[6] Canadian Wood Council (CWC). 2019. Available from: www.cwc.ca [Accessed: December 20, 2018]

[7] Canadian Wood Council (CWC). Wood Design Manual. Ottawa, ON, Canada: Canadian Wood Council; 2017. 1486 p

[8] Gong M, Delahunty S, Rao SZ, Chui YH. Development of a material-efficient finger-joint profile for structural finger-joined lumber. In: Proceedings of the 12th World Conference on Timber Engineering; 16-19 July 2012; Auckland, New Zealand; 2012

[9] The Engineered Wood Association (APA). 2019. Available from: www.apawood.org [Accessed: February 10, 2018]

[10] Dorey AB, Cheng JJR. Glass Fiber Reinforced Glued Laminated Wood Beams. Edmonton, AB, Canada: University of Alberta; 1997. 88 p

[11] Canadian Standards Association (CSA). Engineering Design in Wood. Mississauga, ON, Canada: CSA; 2014. 280 p

[12] Canadian Wood Council (CWC). Innovative Applications of Engineered Wood. Ottawa, ON, Canada: Canadian Wood Council; 2018

[13] Lefebvre D, Richard G. Design and construction of a 160-metre-long wood bridge in Mistissini, Quebec. In: Proceeding of Internationales Holzbau-Forum; 3-5 December 2014, Garmisch-Partenkirchen, Bavaria, Germany; 2014

[14] Binational Softwood Lumber Council (BSLC). Nail-Laminated Timber: US Design & Construction Guide. Ver. 1.0. Minneapolis, MN, USA: StructureCraft; 2017. 142 p

[15] Hong KEM. Structural performance of nail-laminated timber-concrete composite floors [thesis]. Vancouver, Canada: The University of British Columbia; 2017

[16] StructureCraft. 2019. Available from: www.structurecraft.com [Accessed: December 18, 2018]

[17] Haller P, Pannke K. Structural and physical behaviour of nailed laminated timber elements. In: Proceedings of the World Conference of Timber Engineering (WCTE), Vol. 2. 17-20 August 1998; Montreux-Lausanne, Switzerland; 1998. pp. 230-237

[18] StructureCraft. Dowel Laminated Timber—Mass Timber Design Guide, Ver. 3. Abbotsford, BC, Canada: StructureCraft; 2017. 24 p

[19] Mohammad M, Gagnon S, Douglas BK, Podesto L. Introduction to cross laminated timber. Wood Design Focus. 2012;**22**(2):3-12

[20] FPInnovations. CLT Handbook (Canadian Edition). Special Publication: SP-528E. Quebec, QC, Canada: FPInnovations; 2011. 594 p

[21] FPInnovations. CLT Handbook (US Edition). Special Publication: SP-529E. Pointe-Claire, QC, Canada: FPInnovations; 2013. 572 p

[22] Gong M, Niederwestberg J, Chui YH. Development of a lumber-SCL panel product. Report #: WSTC2013-015. Fredericton, NB, Canada: Wood Science and Technology Centre, University of New Brunswick; 2015. 55 p

[23] Niederwestberg J, Zhou JH, Chui YH, Gong M. Shear properties of innovative multi-layer composite laminated panels. In: Proceedings of the 15th World Conference on Timber Engineering; 20-23 August 2018; Seoul, Korea; 2018

[24] Zhou J, Niederwestberg J, Chui YH, Gong M. 2018. Bending properties of innovative multi-layer composite laminated panels. In: Proceedings of the 15th World Conference on Timber Engineering; 20-23 August 2018; Seoul, Korea; 2018

[25] Canadian Wood Council (CWC). Brock Commons Tallwood House—A Case Study. Ottawa, ON, Canada: Canadian Wood Council; 2018

[26] Ramage MH, Burridet H, Busse-Wicher M, Fereday G, et al. The wood from the trees: The use of timber in construction. Renewable and Sustainable Energy Reviews. 2017;**68**:333-359

[27] International Organization for Standardization (ISO). Environmental Management—Life Cycle Assessment—Principles and Framework, ISO 14040. Brussels, Belgium: International Organization for Standardization; 2006

Chapter 3

Structural Design of a Typical American Wood-Framed Single-Family Home

Anthony C. Jellen and Ali M. Memari

Abstract

Light-wood framing construction techniques have been traditionally used in America for the construction of single-family residences. Dimensional wood lumber is readily available and due to its convenient unit dimension can be packaged neatly and transported to work sites by either commercial transport or personal vehicle. The unit pieces of dimensional lumber are light and easily handled once on the work site. Design of light-framed single-family homes is typically conducted by an architect or construction contractor using prescriptive building codes. A structural engineer can assist, if needed, with design items not within the scope of the building code or if alternative design approaches are required. An owner may choose to involve the engineer to improve quality or economy of the home design. Engineers typically become involved with design items such as foundation design, steel framing design, or engineered product specification. In this chapter, the design of a typical light-framed home is discussed. The main structural assemblies are described and subsequently designed using a combination of prescriptive guidance and engineering design.

Keywords: residential, single-family home, wood, light-framing, house

1. Introduction

The prevailing system used for the construction of single-family homes in the USA is platform framing using light wooden dimensional lumber. Structural assemblies such as the roof, floors, and walls are generally constructed with nominal 50.8 mm (2 inch) lumber members ranging in nominal depths from 101.6 to 304.8 mm (4–12 inches) and sheathed with structural wood panels for stability and security, such as oriented strand board (OSB) or plywood.

Wood structural materials are preferred by US homebuilders largely because (1) the US home building industry is mostly familiar with wood framing method, (2) the units of construction (i.e., studs, joists, panels, etc.) are small and easily transportable, and (3) wood-framed structures can be erected without the need for specialized tools or large equipment.

In this chapter, the complete process of designing a typical US residential dwelling using wood-frame systems will be illustrated. The typical US design methodology and basis will be used to accomplish the designs. The International Residential Code (IRC) [1] is the design basis used by most authorities to regulate the design and construction of single-family residences. The following major aspects are discussed in this chapter:

1. Provide introductory material such as the description of the home to be designed, applicable design codes, and external loading assessment for residential structures.

2. Design the home using a wood-framed platform system. The load path will be discussed as well as specific design codes relating to wood-framed structures. The result of specifying and detailing typical structural elements of the home will be specified and details provided.

The scope is limited to the structural design and performance of one single-family residential dwelling. The load-bearing wall systems are the primary components of the building enclosure, and the structural properties of the wall system are only one of many considerations that must be taken into account. While cladding compatibility, thermal performance or the hygrothermal characteristics of a wall system are very important, such aspects are not the focus of this study and will not be discussed.

The home design considered in this study is a two-story regular-shaped home with a basement and attached two-car garage. The floor plan was provided by S&A Homes, which is a midsized homebuilder that builds homes and provides architectural design services to customers in Pennsylvania and West Virginia. The floor plans and drawings for one of their standard home packages are provided in the Appendix. Clients of S&A Homes can select this floor plan from an array of floor plans and make slight variations to it if desired. S&A Homes will then design, detail, and construct the home for the client on the chosen lot, typically one of S&A's own residential developments.

The home plan/style shown in the Appendix is a popular model in S&A's territory and is representative of the size and style of homes desired by the average homebuyer of this decade. The home consists of nearly 214 m^2 (2300 ft^2) of finished floor area with the basement available for finishing if so desired by the prospective homeowner. The floor plan has features typically seen in modern homes. The first floor contains a large kitchen open to the family room with access to both the dining room and the attached two-car garage. The second floor has four bedrooms with the master suite containing its own large bathroom as well as a sitting area and walk-in closet (WIC).

2. Applicable codes and standards

The IRC is the prevailing design code used for the construction of one- or two-family dwellings in the USA. The 2015 IRC [1] is the current adopted code in the State College, PA area, and will be used as the governing design code for this study. In order to construct a single-family dwelling, the homebuilder must first apply to the local code office for a building permit. It is necessary to provide a complete architectural plan set detailing how the builder intends to comply with the requirements of the IRC, along with several other items such as the manual J [2] heat loss-gain calculations for the structure and selection of energy compliance path. The IRC largely provides a prescriptive basis for home design and in many instances is adequate for single-family home design. The envelope and structural components are typically selected by the architect, builder, or homeowner from design tables within the code. If prefabricated engineered components such as I-joists, laminated veneer lumber (LVL) components, or roof trusses are used in design, a structural engineer is required to review their specification and application.

This is typically the extent of a structural engineer's involvement in residential design other than specialized situations not covered by the IRC and occasionally foundation design. If engineered design is necessary in conjunction with the prescriptive standards, then compliance with the 2015 International Building Code (IBC) [3] requirements for those portions of the design is required. Engineers will conduct their analysis based on requirement set forth in the IRC, IBC if necessary, and ASCE 7-10 minimum design loads for buildings and other structures (ASCE 7) [4] [ASCE stands for American Society of Civil Engineers]. The IRC and IBC also permit designers to refer to the 2015 AWC *Wood Frame Construction Manual* (WFCM) [5] for an alternative prescriptive or engineered approach [AWC stands for American Wood Council].

3. External load determination and serviceability requirements

This study will focus on the appropriate residential structural building loads for the State College, PA area, for an example design case. The designs will include only the effects of dead loading, floor live loading, roof live loading, snow loading, and wind loading. Residential structures in ordinary situations are designed to resist both gravity loads and lateral loads. External loading for homes is prescribed in either Chapter 3 of the 2015 IRC or in ASCE 7. ASCE 7 is the standard referenced in the 2015 IRC, and therefore this version will be referenced in this study. Both the IRC and the ASCE 7 will be used to develop the external loads for this study. In addition to the external loads, the serviceability criteria must also be considered. For this design, only live load deflection limits will be considered.

3.1 Gravity loads

The gravity loads are those loads that act in the direction of gravity. The gravity loads of importance for residential structures are dead load (D_L), floor live load (L_L), roof live load (R_L), and snow load (S_L).

3.2 Dead load (D_L)

Dead load is the load that is permanently and continuously applied to a structure. Typically, dead load refers to the self-weight of the material used in construction or a load that is applied in a permanent nature such as a known location of a piece of heavy equipment or a large island in the kitchen. Unless noted otherwise, the S&A Homes dead load criteria will be used for the wood-framed design of this home. These loads are typical for residential design and were largely derived from ASCE 7 Table C3-1. Dead loads are listed in **Tables 1–3**.

3.3 Live load

Live loading is a gravity loading that is temporary or intermittent in nature. The three live loads considered for the design of this home are floor live (L_L), roof live (R_L), and snow load (S_L). The IRC prescribes the minimum uniformly distributed loads that must be used by designers for residential structures. Such minimum loads listed in **Table 4** will be used for this study.

Sub-component	Weight N/m² (lbf/ft²)
Carpet/vinyl	47.9 (1.0)[a]
19.1 mm (¾ in) plywood	114.9 (2.4)
301.6 mm (11 7/8 in) I-joists[b]	91.0 (1.9)
Mechanical allowance	95.8 (2.0)
12.7 mm (½ in) gypsum ceiling	105.3 (2.2)
Total	≈454.9 (10)

[a]*For floor areas known to have ceramic tile floor covering, increase load to 0.96 kN/m² (20 lbf/ft²).*
[b]*Weight is derived from Weyerhaeuser publication #TJ-4000 for 230 or 360 series joists.*

Table 1.
Floor/ceiling assembly weight.

Sub-component	Weight N/m² (lbf/ft²)
Truss framing	95.8 (2.0)
11.1 mm (7/16 in) sheathing	81.4 (1.7)
Asphalt shingles	114.9 (2.4)
228.6 mm 9 in insulation	86.2 (1.8)
12.7 mm (½ in) gypsum board	105.3 (2.2)
Miscellaneous	95.8 (2.0)
Total	≈ 579.4 (12)

[a]*Engineered wood truss roof system.*

Table 2.
Roof assembly weight.[a]

Sub-component	Weight
Exterior wall assembly[a]	526.7 N/m² (11.0 lbf/ft²)
Interior wall assembly[b]	383.0 N/m² (8.0 lbf/ft²)
Plain concrete	22.8 kN/m³ (145 lbf/ft³)
Reinforced concrete	23.6 kN/m³ (150 lbf/ft³)

[a]*2 × 6 wood studs at 406.4 mm (16 inch) O.C. with 12.7 mm (½ inch) gypsum wallboard and vinyl siding.*
[b]*Wood or steel studs with 12.7 mm (½ inch) gypsum wallboard on each side.*

Table 3.
Miscellaneous materials.

3.4 Lateral loading

The only lateral load being considered for this study is the wind loading. In the State College area, seismic loading does not typically control the design of structural components. The procedures in ASCE 7 will be used to determine wind loading, e.g., Chapter 28 Envelope Procedure Part 2 can be used for this structure. Chapter 28 requires that the structure meets the definition of a low-rise, enclosed simple diaphragm building that is regular-shaped in accordance with Section 26.2.

Load description	Weight kN/m^2 (lbf/ft^2)
L_L (sleeping rooms)	1.44 (30.0)
L_L (other)	1.92 (40.0)
L_L (habitable attics)	1.44 (30.0)
L_L (attics w/limited storage)[a,b]	0.96 (20.0)
L_L (Attics w/o limited storage)[c]	0.48 (10.0)
Roof live load	0.77 (16.0)
Design roof snow load[d]	1.44 (30.0)

[a]*Attics defined as the unfinished area between the roof and the ceiling of the floor below.*
[b]*Limited storage refers to non-habitable attic space greater than or equal to 1.07 m (42 inch).*
[c]*Add to attic space less than 1.07 m (42 inch).*
[d]*Based on State College area prescriptive requirements. Applied on the horizontal projection rather than along the slope.*

Table 4.
Minimum uniformly distributed live loads.

Parameter	Description
Risk category	II
Basic wind speed (V)	51 m/s 115 mph
Exposure category	B
Topographic factor (K_{zt})	1.0
Mean roof height	7.0 m (23 ft)
Adjustment factor (λ)	1.0
Roof pitch	30 degrees

Table 5.
Wind load parameters.

Zones	Case 1	Case 2	Minimum
A	1.13 (23.6)	1.13 (23.6)	0.77 (16)
B	0.77 (16.1)	0.77 (16.1)	0.38 (8)
C	0.90 (18.8)	0.90 (18.8)	0.77 (16)
D	0.62 (12.9)	0.62 (12.9)	0.38 (8)
E	0.09 (1.8)	0.44 (9.1)	0
F	−0.68 (−14.3)	−0.34 (−7.1)	0
G	0.03 (0.6)	0.38 (7.9)	0
H	−0.59 (−12.3)	−0.24 (−5.0)	0
E_{OH}	−0.40 (−8.3)	−0.40 (−8.3)	0
G_{OH}	−0.45 (−9.5)	−0.45 (−9.5)	0

Values in kN/m^2 (lbf/ft^2).

Table 6.
Simplified design wind pressure (Ps) case A θ = 30.

The wind loads calculated in **Table 6** are based on the parameters listed in **Table 5** and in accordance with **Figure 1**. The simplified design wind pressure magnitudes in **Tables 6** and 7 include both windward and leeward pressures. The combined pressure will be applied to only the windward side of the structure. For this design, two load cases must be evaluated because the roof pitch is between 25 and 30 degrees. Additionally, these two cases must be compared to the minimum load case described in ASCE 7 Section 28.6.4. The case that produces the larger load effect will be used for design of structural members.

3.5 Serviceability criteria

The main serviceability criterion considered in the design of residential homes is deflection. The IRC prescribes the maximum allowable deflection of structural

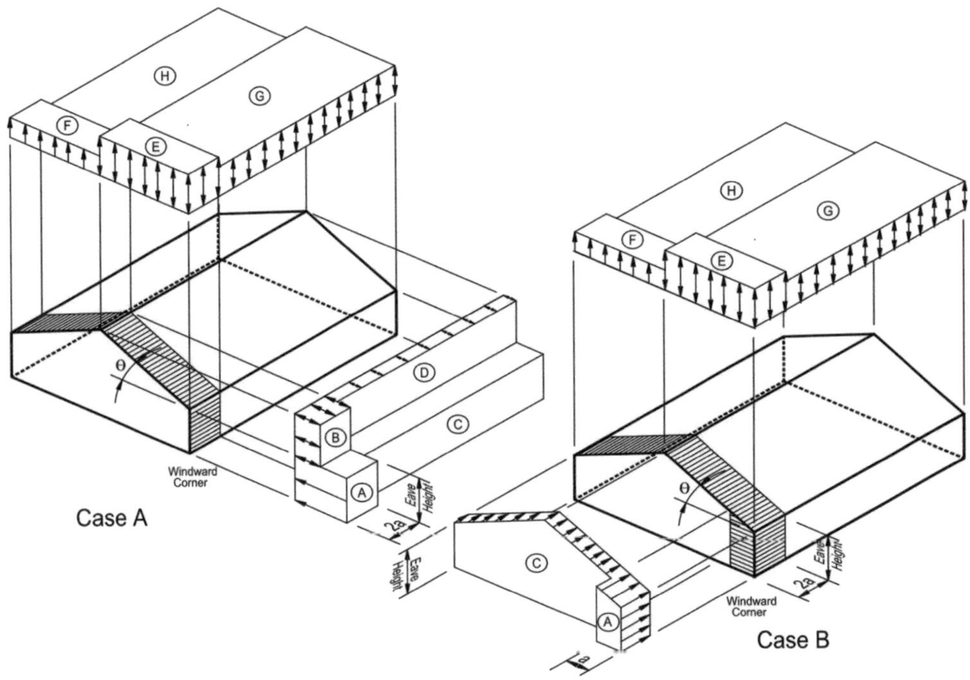

Figure 1.
ASCE 7-10 Chapter 28 wind loading designation (with permission from the ASCE).

Zones	Case 1	Minimum
A	1.01 (21.0)	0.77 (16)
C	0.67 (13.9)	0.77 (16)
E	−1.21 (−25.2)	0
F	−0.68 (−14.3)	0
G	−0.68 (−14.3)	0
H	−0.53 (−11.1)	0

Values in kN/m² (lbf/ft²).

Table 7.
Simplified design wind pressure (Ps) case B θ = 0.

Sub-component	Span ratio
Interior walls and partitions	Height/180
Floors and plaster ceilings[a,b]	Length/360
All other structural members	Length/240
Exterior walls—brittle finish	Length/240

[a]*Limit floor beam deflection to 12.7 mm (½ inch).*
[b]*Limit I-joist deflection ratio to length/480.*

Table 8.
Live load maximum deflection limits.

members and assemblies. Excessive deflections can cause problems for the occupants and potentially damage to nonstructural components such as cladding or fenestration. Excessive interior floor deflections are generally noticed in the form of floor vibration or "spongy" floors. Excessive deflection of roof members can lead to ponding and ultimately moisture issues or overloading of structural members. A portion of Table R301.7 from the IRC that prescribes residential deflection limits is reproduced below in **Table 8**.

3.6 Combination of loads

Both allowable stress design (ASD) and load resistance and factor design (LRFD) load combinations will be utilized for different aspects of the home structural design. For example, the ASD approach will be used for wood design, whereas the LRFD approach will be used for concrete foundation design. Approaches for the designs will be discussed as appropriate. The load combinations that will be used for design are listed below and are reproduced from ASCE 7.

3.6.1 ASD load combinations

1. D
2. D + L
3. D + (Lr or S or R)
4. D + 0.75L + 0.75(Lr or S or R)
5. D + (0.6W or 0.7E)
6. D + 0.75L + 0.75(0.6W) + 0.75(Lr or S or R)
7. 0.6D + 0.6W

3.6.2 LRFD load combinations

1. 1.4D
2. 1.2D + 1.6L + 0.5(Lr or S or R)
3. 1.2D + 1.6(Lr or S or R) + (L or 0.5W)

4.1.2D + 1.0W + L + 0.5(Lr or S or R)

5. 0.9D + 1.0W

In the above load combination, the notation is defined as follows: D for dead load, L for live load, Lr for roof live load, S for snow load, R for rain load, and W for wind load.

4. Design of residence

Wood is the most popular material used in the USA for the construction of single-family dwellings. An example of residential framing can be seen below in **Figure 2** [6]. Framing lumber is easily obtained in most locations. The units of construction can be easily transported by contractors or homeowners without the need for specialized equipment. Additionally, the erection of a wood-framed structural system is familiar to most and does not require excessive amounts of

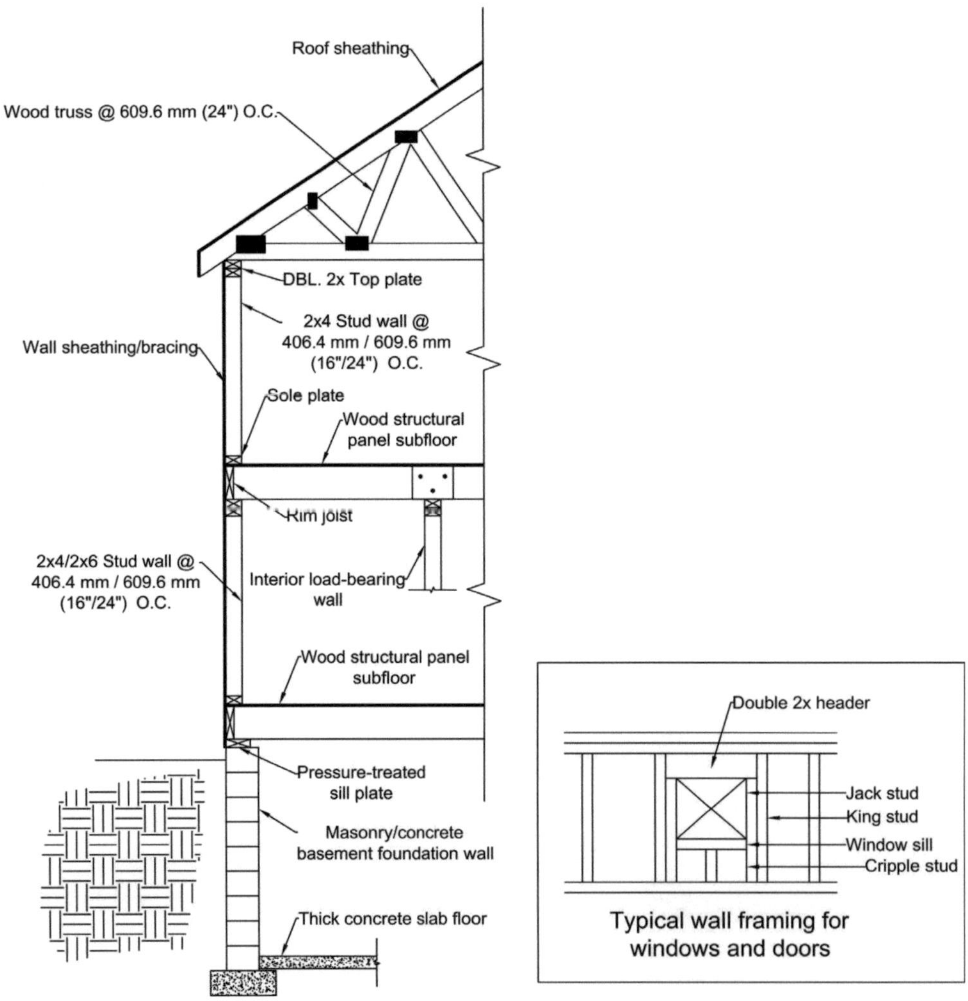

Figure 2.
Section view of typical residential wood-framed home. Note: in this figure, a small rectangle with x inside indicates the cross section of wood member, and DBL stands for double.

specialized knowledge or tools. Lastly, wood-framed construction has been well documented in the USA, and many design aids are available.

As noted before, much of the wood-framed structural design can be accomplished using design aids. The design professional will typically use these design aids to the greatest extent possible and then perform structural analysis and design for any item that is beyond the scope of the design aids. This is the approach that will be used for this study. The design drawings are shown in the Appendix. The associated detailed calculation is not provided due to space limitation; only the necessary results will be mentioned.

4.1 External load transfer (load path)

External loads must be transmitted to ground through the structural system of the building. Two main systems are needed to accomplish this transfer properly: gravity system and the main wind force resisting system (MWFRS). The gravity system transmits the vertical loads through a system of trusses, joists, and beams to foundation, which in turn transmits the load to ground, while the MWFRS transfers lateral wind load to foundation through a system of shear walls and flexible diaphragms. It is important to recognize that the ground must be properly prepared and evaluated to ensure good load transfer. Typically, foundations are placed on virgin soil or engineered (compacted) fill. All organic materials should be removed along with excessive amounts of water.

4.2 Gravity system design

The gravity system in this home starts at the roof and ends in the soil. Vertical loads must have a continuous path to the ground. Generally, the gravity system in this example consists of OSB sheathing, engineered roof trusses, load-bearing stud walls, dimensional lumber headers, engineered I-joist floor system, engineered wood beams, structural steel girders, and a concrete foundation.

4.3 Roof sheathing

The OSB roof sheathing, as illustrated in **Figure 3**, serves to transfer gravity load (i.e., dead, live, and snow loads) and wind suction to roof framing members. The roof sheathing also transfers the lateral wind loading through diaphragm action to the structure. Attachment requirements of the sheathing to roof trusses are governed by the greater of the wind uplift force or the shear transfer requirement of the connection.

According to IRC Table R503.2.1.1(1), 11.1 mm (7/16 inch) roof sheathing (24/16 span rating) is acceptable for this example. The sheathing can be used with or without edge support at 609.6 mm (24 inch) spans with an allowable live load of 1.92 kN/m^2 (40 lbf/ft^2), and a total allowable load of 2.39 kN/m^2 (50 lbf/ft^2), which is less than the 1.44 kN/m^2 (30 lbf/ft^2) snow loading plus 0.57 kN/m^2 (12 lbf/ft^2) roof dead load. It may be possible to use 9.5 mm (3/8 inch) sheathing, but 11.1 mm (7/16 inch) thickness is more readily available and common in the locale. In this example, the sheathing will be specified with panel edge clip support. According to IRC Table R602.3(1), the sheathing is required to be attached to the truss framing with 63.5 mm (2½ inch) 8D common nails spaced at 152.4 mm (6 inch) on center (O.C.) around the edges of the panel and 304.8 mm (12 inch) O.C. at intermediate supports (field). Note that the gable end sheathing connections must be spaced at 152.4 mm (6 inch) O.C. at both the perimeter and intermediate locations.

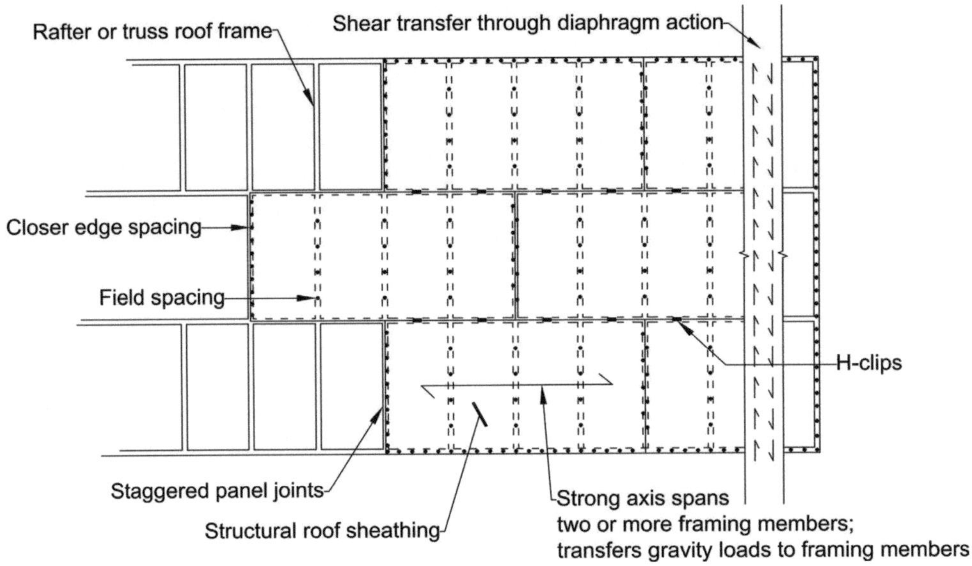

Figure 3.
Roof sheathing illustration.

4.4 Engineered roof trusses

Prefabricated trusses are intended to be used on this residence and required engineering design by the manufacturer. Wood roof trusses must be designed in accordance with IRC Section R802.10. A designer or architect will typically draw the shape of the roof system, and then the truss designer will design the truss system to fit the concept. Typically, it is the responsibility of the home designer to ensure that the gravity and lateral loads from the trusses are properly transferred to the wall below. This involves specifying the connection to wall system below. When the truss drawings are received by the home designer, the loads to the structure, based on the analysis conducted by the truss designer, are typically listed on the engineered truss plans. The designer would use these loads for design. For the example case presented here, however, a set of detailed truss drawings are not available. The assumed loadings described earlier will be used for design. This is typical of an initial home design. A designer will use their assumptions and then verify such assumptions when the final truss plans are received.

4.5 Exterior walls

The gravity load-bearing elements of the wall system presented here are the 2 × 6 dimensional lumber studs and the top and bottom plates (or sole plate). See **Figure 2** for the location of the top and bottom plates. The 2 × 6 designation refers to a wood framing member with a nominal 50.8 mm (2 inch) width and a 152.4 mm (6 inch) depth. The actual measurements of the member are approximately 38.1 mm (1½ inch) wide and 139.7 mm (5½ inch) deep. The top and bottom plates serve to transfer both gravity and lateral loads between floors. The top plate serves three purposes: (1) a chord for the MWFRS, (2) a strut between shear panels in a wall line, and (3) a means to transfer gravity loads to the stud from the joists and trusses.

According to IRC Table 602.3(5), 2 × 6 studs can be used at 609.6 mm (24 inch) O.C.; however, it is more typical for the studs to be spaced at 406.4 mm (16 inch)

O.C. The advantage of this is that when using a double 2 × 6 top plate, the joists or trusses that bear on the wall do not have to bear directly on the stud. If using a single top plate or studs spaced at 609.6 mm (24 inch) O.C., then the joists or trusses must either be directly above the stud or within 25.4 mm (1 inch) of the stud according to IRC Section R602.3.2. It is possible to use 2 × 4 studs spaced at 406.4 mm (16 inch) O.C., but this is not common because of the popularity of using fiberglass batts to meet the International Energy Conservation Code (IECC) [7] envelope insulation requirements. The connections between the studs and the plates are according to IRC Table 603.2(1). The connections are typically nails, and the nail sizes vary between 8D and 16D based on the detail.

4.6 Headers within wall system

Structural header members are used to create openings in a load-bearing wall assembly for fenestration (windows and doors) as shown in **Figure 2**. Dimensional lumber headers are preferred by designers when loading is low. Often times when point loading is present on a header or spans are large, an engineered lumber header, such as an LVL, may become cost-effective. An example of a typical LVL is shown in **Figure 4**. LVLs are also often used in wall systems when smaller depth members are required due to space constraints.

When specifying headers, the designer may choose to specify larger headers in some locations for consistency sake. By minimizing the amount of different beam sizes on the plan, the designer reduces the risk of misplacement of headers. As in the case of the roof sheathing, it may also turn out that some beam sizes may be more readily available, and therefore larger sections may be more economical. For example, a two-ply 2 × 8 beam, with a demand capacity ratio of 0.944 controlled by bearing, is adequate for BM3, but because the entire back wall on the first floor is composed of two-ply 2 × 10 headers and all the other headers in the building are 2 × 6's, it makes sense just to specify a two-ply 2 × 10 beam for this location as well. This eliminates the need to have another beam size on site and provides for the opportunity to use trim pieces from a different header cut to make up this short beam.

4.7 Above-grade floor system

In this home design, an engineered floor system will be used. As shown in **Figure 4**, I-joists have become popular and cost-effective in the residential home construction market. I-joists have several advantages over dimensional lumber

Figure 4.
Typical I-joist and LVL (courtesy Timber Rock Homes).

joists, one of which is a greater span-to-depth ratio. This allows for shallower floor assemblies, longer spans, and higher ceilings. I-joists are generally more stable than dimensional lumber. This almost eliminates the need for bridging in a floor system and ensures consistency of engineering properties.

An I-joist floor system is an engineered product. Typically, a designer will send their floor plan along with preliminary input from the designer to the I-joist manufacturer. The manufacturer will then design the floor system according to the requests of the homeowner and designer. Live load deflections are often limited to L/480 (beam span/480). Because longer spans can be achieved by using an I-joist product, the chances of floor vibration occurring increase, but can be controlled, as designers will often restrict deflection to L/480.

It is common for designers to use span tables to select an initial floor joist size. This will provide a fairly accurate estimate and allow the designer to select a floor assembly depth. The improved stability and increased stiffness of I-joists allow designers to consider larger spacing for the floor joists. It is common to specify I-joists at 487.7 mm (19.2 inch) O.C., whereas it was generally common in the past to specify dimensional lumber joists at 406.4 mm (16 inch). Additionally, lumber joists are only available in certain lengths. This made the need for a splice at an internal bearing wall or beam a very common occurrence. The length of I-joists is generally only limited by transportation and site restrictions. An I-joist package will typically arrive at the site precut and ready to be installed with minimal modification.

As in the case of roof sheathing, floor sheathing serves two purposes. First, it acts in the gravity system to distribute floor loads to the joists. Secondly, it is the primary shear resisting component in the floor diaphragm, which will be discussed subsequently. Typically, the gravity loads govern the thickness choice of subflooring, and the shear requirements dictate connection to joists [8].

Once again IRC Table R503.2.1.1(1) will be used to size the sheathing. In this case, the sheathing will serve as both the underlayment and the subflooring. From the table, either 18.3 mm (23/32 inch) or 19.1 mm (3/4 inch) tongue and groove oriented strand board (OSB) sheathing would be appropriate, whichever is more cost-effective and readily available. It is possible that 15.1 mm (19/32 inch) or 15.9 mm (5/8 inch) sheathing could be used, but spans are restricted to 508 mm (20 inch). Although the joists will be specified at 487.7 mm (19.2 inch), which is less than the limit, it is likely that at least a few joists within the floor system will need to be spaced greater than 508 mm (20 inch). An example is when joist bays are used for heating, ventilating, and air conditioning (HVAC) ductwork, the joists are often spread in those locations to 609.6 mm (24 inch). In this instance, the thinner sheathing would be inadequate. IRC Table 602.3(1) specifies attachment of the sheathing to joists with a 50.8 mm (2 inch) 6D deformed nail or a 63.5 mm (2½ inch) 8D common nails spaced at 152.4 mm (6 inch) O.C. around sheathing edges and 304.8 mm (12 inch) O.C. for intermediate field spacing.

Joists for this project are selected from the Trus Joist #TJ4000 specifier's guide [9]. From the span tables within the guide, TJI110 301.6 mm (11 7/8 inch) joists are adequate for both the first and second floors of this residence. The maximum span in the home is approximately 4.70 m (15 foot–5 inch). The TJI110 301.6 mm (11 7/8 inch) joist can span a maximum of 4.90 m (16 foot–1 inch) considering L/480 deflection limit, 1.92 kN/m^2 (40 lbf/ft^2) live load, and a 0.96 kN/m^2 (20 lbf/ft^2) dead load. The TJI 28.6 mm (1 1/8 inch) engineered rim board will be used for the perimeter of the floor system. The rim board serves to transfer compressive and shear loads from the exterior walls above to foundation below. It also acts to enclose the perimeter of the floor system. Typically, joists are toenailed to sill plates at ends and nailing plates at intermediate points. Metal hardware such as that made by USP [10] or Simpson Strong Tie [11] is used to make any flush beam-to-beam or joist-to-beam connections within the floor system. An example would be the stair trimmer detail shown in **Figure 5**.

A double joist or LVL product can be used to function as stair trimmers in an engineered floor system. When loads are low, double joists are economical, but as loading and span increase, an LVL is sometimes needed. LVLs are sometimes used because the installation is cleaner looking and easier to finish than double joists. Double joists often require padding at connections and sometimes bearing, which is usually OSB, to compensate for the space between the web and flanges. LVLs are conveniently made in the same depths as I-joists, which makes it easy to use within the floor systems.

A benefit of using I-joists over dimensional lumber is that it is easier to put holes through the joists for mechanical runs. Most I-joist manufacturers will have predetermined locations or precut holes in the joists where mechanical penetrations are anticipated. Some guidance is typically specified in the manufacturer literature. Holes in dimensional lumber typically require structural analysis and stress evaluation as they become large relative to the depth of the joist or beam.

4.8 Girder sizing

For this example home design, a central steel girder will be used to collect the floor loads and transfer to pad footings in the center of the basement. It is common for designers to use either steel girders or manufactured lumber girders in homes today. These types of girders are much stronger than dimensional lumber beams and are necessary in many instances because of the longer allowable engineered I-joist spans and homeowner request for open basement floor plans. Both manufactured lumber girders and steel girders must be either specified or the design reviewed by a professional engineer.

Steel girders are often chosen over manufactured lumber girders when girder spans are long, head room in the basement is a premium, or steel is readily available. For this particular builder, the head room in the basement is important because they like to advertise their homes with basements that can be finished in the future.

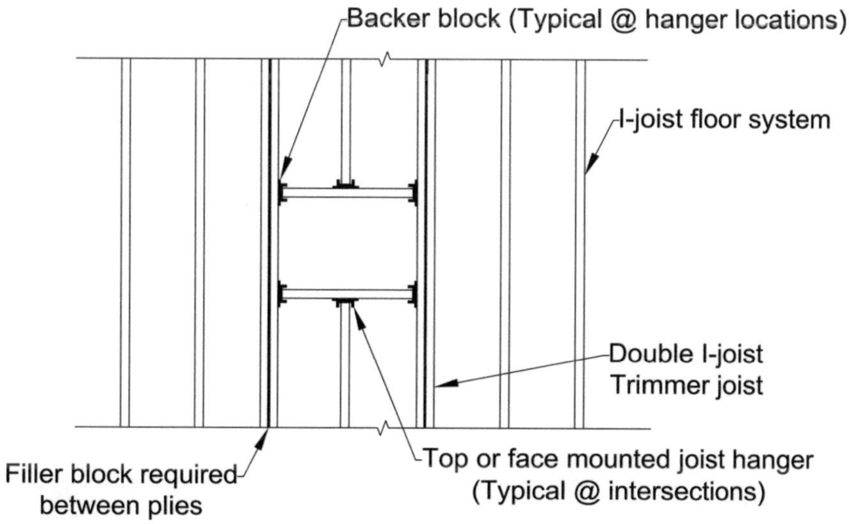

Figure 5.
Stair trimmer detail.

A W8x18 girder works well for them because it's a shallow beam and the flange width is small enough that the beam can fit in a 2 × 6 wall making the girder unnoticeable if the basement is ever finished.

A W8x18 steel girder, with a design moment capacity of 86.5 kN-m (63.8 kip-ft), is more than adequate to resist the internal moment of 31.5 kN-m (23.2 kip-ft) for the controlling load case. It is possible that a smaller girder could have been used, but W8x18 is the minimum size the builder will use. Small sizes tend to have stability issues and can be susceptible to local buckling problems caused by larger point loads. In addition, this is a readily available steel section from the builder's steel supplier.

The design of residential girders involves assumptions regarding the bracing of the beam. The American Institute of Steel Construction (*AISC*) *Steel Construction Manual 14th Ed.(SCM)* in Chapter B3.6, F1 (2) [12] and Appendix 6.3 all require that girders are restrained against rotation about their longitudinal axis at the points of support unless it can be shown that the restraint is not required. The amount of restraint provided by the adjustable column, which is typically four bolts through the bottom flange, may need a detailed analysis because of the slenderness of the columns.

Steel girders in most residential cases are usually ordered in a single length if possible to avoid splices and therefore are continuous over their intermediate supports. Negative moment occurs at the intermediate supports, which puts the bottom flanges in compression in those regions.

If it is assumed that the columns do not provide adequate bottom flange support, then these negative moment regions would be destabilizing, and since inflection points are not typically recognized as a brace point (SCM Appendix 6.3), the unbraced length would have to be taken as the entire beam length of 11.0 m (36 feet), which would require a very large section. Additionally, if no compression flange bracing is assumed at the supports, then the beam fails the concentrated load check in SCM J10.4 for web sidesway buckling. Section J10.4 requires the supports to be adequately braced under these circumstances.

If it is assumed that the column is braced against rotation at the supports by either assuming the column connection is adequate or providing additional bottom flange support, then the unbraced length reduces to the distance between the columns, which in this case is 9′-0″ and the beam passes both strength and concentrated load checks.

Also restraint against rotation should be provided at the ends of the beams, which are seated in the beam pockets. Typically, beam pockets in the concrete wall are oversized to facilitate easy installation of the beams. This creates the opportunity for twisting. SCM Section J10.7 requires all unframed girder ends to have a pair of transverse stiffeners if unrestrained. In this case, a better idea would be to grout the pocket as shown in **Figure 6**, or provide some type of shim, after installation to restrain the end against rotation. It should be noted that the required moisture management and thermal envelope components are not shown for clarity in the figure.

Another consideration for girder sizing is live load pattern loading. Since the girder is a continuous beam having multiple spans, ASCE 7 Section 4.3.3 requires the consideration of pattern loading. In this case, it turns out that applying live loading to spans 1, 2, and 4 only produced the largest internal moment of 31.5 kN-m (23.3 kip-ft) in the beam. **Figure 7** shows the moment diagram for the controlling load combination and the spans that were loaded to produce it.

Pattern loads are considered in the structural analysis software package Enercalc that was used for beam design. Enercalc runs all permutations of live load application and reports the worst-case scenario in envelope format. Data for individual permutations is not able to be extracted. For this example, a separate check was

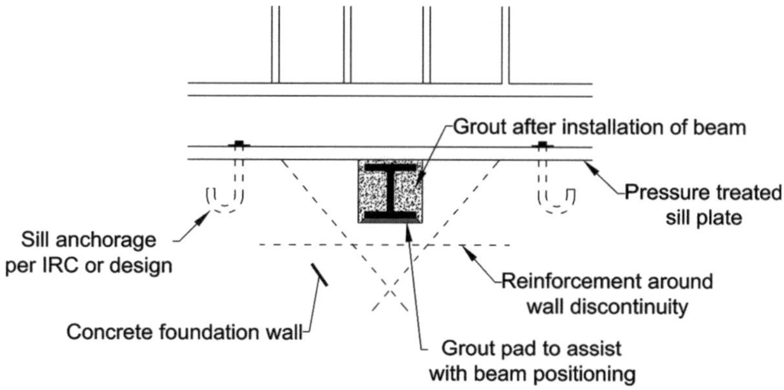

Figure 6.
Steel girder beam pocket detailing.

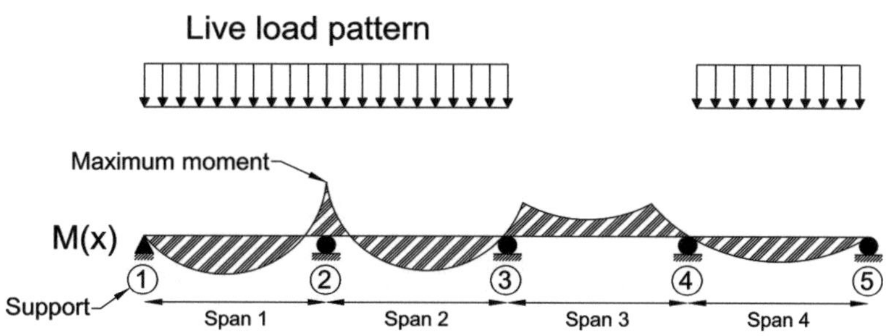

Figure 7.
Moment diagram showing maximum internal moment over support 2.

made using Computers and Structures, Inc. (CSI) SAP2000 finite element modeling software to verify the results of Enercalc and determine the controlling permutation. Results were within 1% of each other between the two analysis packages.

Pattern loading was significant in this example. If only the full intensity live load application was to be considered, then the design moment would have been underestimated by approximately 5%, and the support reactions would have been underestimated by approximately 5% at supports 2, 4, and 12% at support 3. If ignored, this could have led to the undersizing of both adjustable column and pad footing.

4.9 Adjustable columns

Adjustable columns are generally used in residential construction as intermediate supports for basement girders. Adjustable columns are readily available at almost any hardware stores and can be adjusted in height to match site conditions by the contractor. **Figure 8** shows an example of typical adjustable columns. The maximum loading, as reported by the manufacturer, is a factored allowable ASD load capacity (Ra). Reactions determined by ASD load combination can be used to directly size the column from the manufacturers testing data. For this particular home design case, the maximum ASD girder reaction is 80.5 kN (18.1 kip). According to the manufacturers data, an 88.9-mm (3½ inch) and 2.31-mm-thick (11 gauge) column with a height between 2.21 m (7 foot–3 inch) and 2.31 m (7 foot–7 inch) has an allowable load of 95.6 kN (21.5 kip), which is greater than the maximum column axial demand of 80.5 kN (18.1 kip). All three columns

Figure 8.
Typical adjustable column.

will be specified for this maximum loading. This will decrease the chances of misplacing columns.

4.10 Foundation design

A combination of components are used to transfer load from the above-grade portion of the home to the ground. In this home, concrete walls supported by concrete strip footings are used to support the exterior walls and resist lateral earth pressure. Interior loads are transferred by the intermediate girder through columns to concrete pad footings. It is common practice in residential design to specify the foundation walls prescriptively but design the footings. This is the approach that is taken for this study. The American Concrete Institute (ACI) 332-08 [13] and ACI 318-14 [14] are used as references for this design. These documents are adopted by the 2015 IRC and often lead to more economical designs when compared to the requirements of the IRC.

4.11 Foundation walls

Based on soil categorization, the ACI provides prescriptive foundation sizing tables in Appendix A of ACI 332, which are usually appropriate for most situations. For most residential designs, geotechnical exploration and lab testing are cost prohibitive, and therefore soil pressures must be assumed. ASCE 7 provides design lateral soil load that can be used in the absence of site-specific geotechnical information.

For this design, the equivalent soil pressure will be estimated at 2.15 kn/m^2 per linear meter (45 lbf/ft^2 per linear foot). According to ASCE 7 Table 3.2.1, this is representative of a type GC soil (unified soil classification), which is described as a clayey gravel, poorly graded, gravel, and sand mix. Assuming horizontal backfill and a vertical foundation wall, this is roughly equivalent to 19.6 kN/m^3 (125 lbf/ft^3) soil with an internal friction angle of 28 degrees [15].

According to ACI 332 **Table 9**, 21 MPa (3000 psi) is the minimum required compressive strength for foundation walls in the severe weather probability category. Because the concrete will be exposed to weathering, it must be air entrained, having an air content of 6% plus or minus 1.5%.

The concrete foundation wall for the main structure in this example has an unsupported height of 2.44 m (8 foot) and will be subjected to approximately 2.13 m (7 foot) of unsupported backfill when in service (**Figure 9a**). For this situation, considering reinforcing bars with a yield strength of 420 MPa (60,000 psi), ACI 332 Table A.4 allows for the use of a plain concrete (no vertical reinforcing needed) 203.2 mm (8-inch)-thick foundation wall. To minimize shrinkage cracking, however, ACI 332 requires the use of three continuous horizontal bars in the wall. One must be placed within 609.6 mm (24 inch) of the top, one within 609.6 mm (24 inch) of the bottom, and the last one in between the other

Walls	Length required	Length provided	Method
First floor			
N	4.24 (167)	8.23 (324)	CS-WSP
S	3.40 (134)	3.66 (144)	WSP
E	3.20 (126)	3.66 (144)[a]	WSP
W	3.20 (126)	3.66 (144)	WSP
Second floor			
N	1.83 (72)	2.44 (96)	WSP
S	1.83 (72)	2.44 (96)	WSP
E	1.52 (60)	3.66 (144)	WSP
W	1.52 (60)	3.66 (144)	WSP
Garage[b]			
N	1.27 (50)	2.44 (96)	WSP
E	1.32 (52)	1.37 (54)	WSP
W	1.32 (52)	2.44 (96)	WSP

[a]*For WSP methods panel lengths between 0.914 and 1.22 m (36 and 48 inches) are allowed but must be adjusted per IRC Table 602.10.3.*
[b]*The required bracing for the garage/main house common wall will be added directly to the first floor north wall.*

Table 9.
Wall bracing. Values in meters (inches).

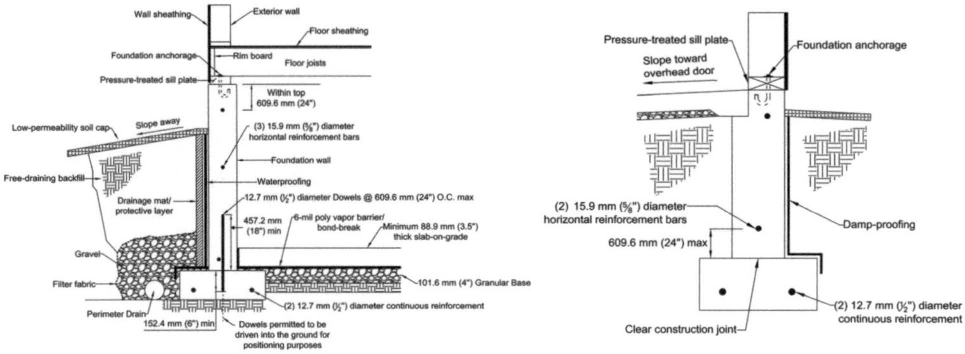

Figure 9.
(a) Typical basement wall and (b) typical garage frost wall.

two. ACI 332 also prescribes 12.7 mm diameter (½ inch) dowel rods at a maximum of 609.6 mm (24 inch) O.C. or a keyway to be provided in this instance since unbalanced backfill height exceeds 1.22 m (4 foot).

The garage wall foundation walls are all 0.91 m (3 feet) in height and have no unbalanced backfill. According to ACI 332, 203.2 mm (8 inch) plain concrete walls are adequate. No vertical reinforcing is necessary, but horizontal reinforcing is still required (**Figure 9b**). The wall height is less than 1.83 m (6 feet), which requires only two 12.7 mm diameter (½ inch) reinforcing bars, one within the top 609.6 mm (24 inch) of the wall height and the other within the bottom 609.6 mm (24 inch) of the wall height. Because the unbalanced backfill is less than 1.22 m (4 feet), Section 6.3.4 allows for the use of a clean construction joint versus dowel rods.

4.12 Wall strip footings

Continuous strip footings will be used to support the exterior foundation walls. The wall footings will be designed (as opposed to prescriptive). No soil testing data is available, so the IRC minimum of 71.8 kN/m^2 (1500 lbf/ft^2) prescribed in Table R401.4.1 will be used for design. The assumption will be made that the footings are not exposed to weathering; therefore, ACI 332 prescribes 17 MPa (2500 psi) minimum compressive strength for the concrete.

For this example, it will be assumed that the load from the exterior wall will act concentrically on the footing. In other words, the footings will be designed for uniform pressure only, and no imbalanced soil pressure due to the presence of a moment will be considered. This is a reasonable assumption because basement walls are typically restrained from translation at the top and bottom by the first floor assembly and the basement slab, respectively. The presence of this restraint allows walls to be designed as a vertical beam with pinned ends (no moment transfer). In addition, the opposing soil exterior lateral loading tends to offset the small amounts of eccentricity created by above-grade wall offsets, so in practice the effects of above-grade wall offsets are generally ignored for wall footing design. **Figure 10** shows an illustration of the analytical model for a typical residential basement wall.

Residential wall footings are typically specified in depths of 152.4 mm (6 inch), 203.2 mm (8 inch), or 254 mm (10 inch), and widths are generally varied in 50.8 mm (2 inch), 76.2 mm (3 inch), or 152.4 mm (6 inch) increments. Both the IRC and ACI 332 allow for the use of 152.4 mm (6-inch)-thick footings (assuming adequate strength), but the developer in this case prefers to use 203.2 mm (8-inch)-thick footings. This allows for some additional safety precaution when plain

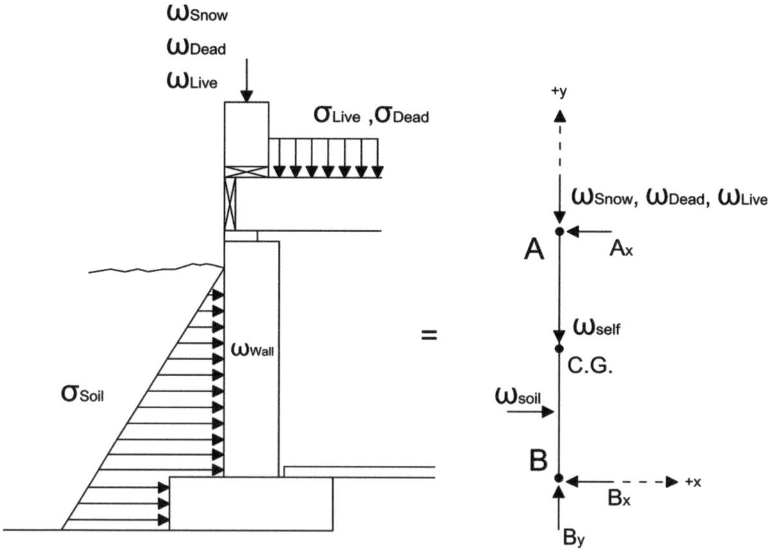

Figure 10.
Free body diagram of a basement wall. Note: the arrows show loads, and small rectangle with x inside indicates the cross section of wood member.

concrete footings are used. When specifying footing widths, this particular developer prefers to use even dimensions in 50.8 mm (2 inch) increments.

In this example, the wall footing design is split into three segments, the main load-bearing walls of the east and west (perpendicular to joist and truss spans), the gable end walls, and the garage walls. Wall footings were designed as plain concrete strip footings according to the requirements of ACI 318, considering the increased modulus of rupture allowed by ACI 332 Chapter 7. Soil bearing pressure controlled all designs. With a soil bearing pressure of approximately 67 kN/m^2 (1400 lbf/ft^2), the bearing walls required 203.2 × 457.2 mm (8 inch by 18 inch) footings. The gable end wall footings and garage footing were able to be reduced to 203.2 × 406.4 mm (8 inch by 16 inch). The wall region beneath the supporting columns for the garage door header controlled the design. Considering ASD load combination 4 and a point load distribution angle of 45 degrees within the concrete wall, the soil pressure beneath the column would be approximately 67 kN/m^2 (1400 lbf/ft^2) as well.

The footings were designed as plain concrete footings. Plain concrete footings are the most economical because of the absence of the steel reinforcing cost. Some developers are comfortable relying on the unreinforced concrete footing to maintain its integrity over the service life of the building, but some prefer to add light reinforcing to help prevent cracking due to unexpected soil discontinuities. ACI 332 Section 6.2.4.1 prescribes the use of two 12.7 mm diameter (½ inch) bars for locations with discontinuities less than 914.4 mm (36 inch) in length.

4.13 Isolated pad footings

Isolated pad footings are typically used to transfer vertical gravity load from interior columns in the basement. In this case, there are three pad footings required to support the interior central steel girder. Interior pad footings are not subjected to weathering, so 17 MPa (2500 psi) concrete compressive strength is adequate. The default value of 71.8 kN/m^2 (1500 lbf/ft^2) is used for the soil bearing capacity, as in the strip footing design.

Reinforced square concrete footings were selected as appropriate for this application. Plain concrete pad footings are sometimes adequate for smaller footings with plan dimension of 609.6 mm (24 inch) or 762 mm (30 inch) square but typically require reinforcement as the plan dimensions of the footing increases. In this case, three 1219.2 mm (4 foot) square footings using four 15.9 mm (5/8 inch) diameter bars in each directions were required. Considering LRFD combination 2, two-way shear (punching shear) with a demand/capacity ratio of 1.30 was the controlling failure mechanism for the concrete footing and required an increase in footing depth from 203.2 mm (8 inch) to 254 mm (10 inch). This reduced the demand/capacity ratio to the acceptable level of 0.698.

4.14 MWFRS design

The typical residential MWFRS system is composed of a system of flexible diaphragms and shear walls. As shown in **Figure 11a**, wind load is transferred from exterior walls perpendicular to the wind direction to structural wood panels, typically OSB or plywood, attached to roof or floor framing. The flexible roof or floor diaphragms, as shown in **Figure 11b**, act similar to a deep beam and distribute the wind load as reactions to the exterior walls parallel to the wind loading (**Figure 11c**) and distribute to the stiff structural shear panels within those walls by direct diaphragm connection or strutting.

The structural wall panels, as shown in **Figure 11d**, provide the necessary shear resistance and transmit the loads vertically (overturning tension and compression

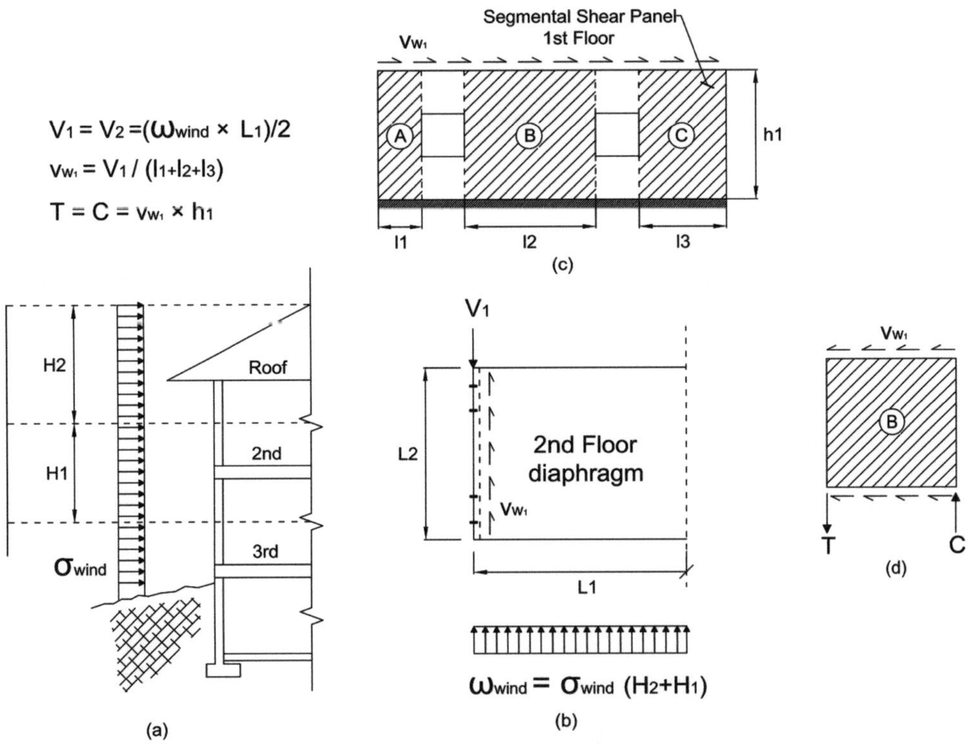

Figure 11.
(a) Wind pressure distributed through external walls to flexible diaphragm. (b) Flexible diaphragm distributes load to parallel walls. (c) An example of a segmental shear wall load distribution approach. (d) Shear wall segment resolution of overturning forces. Note: in this figure, the following notation is used: V for shear force, T for tension force, C for compression force, l and L for Span length, h and H for height, σ for wind pressure, ω for wind load per unit length, and Vw for shear per unit length.

loads at the corners of each structural panel) to the foundation though a system of hold-downs and connections.

Typically, the panels are specified by design aids such as the IRC or the Wood Frame Construction Manual (WFCM). When using the IRC approach, the prescribed nailed connections are assumed to be adequate to transfer the overturning shear forces shown in **Figure 11** to the foundation. If an engineered design or the WFCM prescriptive approach is used to specify shear wall panels, then structural connectors must be specified to transfer these overturning forces. The connection system must have an identifiable load path to the foundation. For this reason, most residential designers use the IRC to specify shear panels and their fastening system. When using a wood truss system as part of the roof diaphragm, such as the one in this home design example, structural connectors are typically specified to transfer the horizontal shear loads and uplift loads resulting from the roof wind loading.

The loads from the shear wall panels and floor diaphragm are transferred to the sole plate by nailed connections and sometimes structural connectors if necessary. The sole plate is attached to the foundation wall with cast-in-place anchors such as J-bolts or post-installed anchorage that must be drilled after the concrete has had time to cure, such as expansion anchors, epoxy anchorage, or screw type. With a prescriptive approach, the prescribed anchor bolts are assumed to adequately transfer both the overturning actions and horizontal actions generated by the wind.

4.15 Overturning and sliding analysis

It's generally good practice to review the whole structure for stability under wind loading and then design the individual components of the lateral force resisting system as required. An overturning and sliding analysis is conducted to determine the required strength of the connections between main assemblies such as the roof-to-wall connections, floor-to-wall connections, and the above-grade building-to-foundation connections.

Many times, homes have attached garages where the garage is not integral to the main living space, such as the one in this example. The garage and the main building can be somewhat treated as separate buildings for the purposes of MWFRS design. The garage can sometimes help resist main building wind loading as long as the wall offsets are not too large; otherwise they must be treated completely separately as far as wall bracing goes. In the east-west direction, the common north wall between the garage and the main structure is generally treated as an exterior wall, and bracing will be prescriptively specified as such, which will act to transfer load from both the garage and the main building.

ASCE 7 Figure 28.6.1 cases A and B were used to determine the magnitude of wind forces applied to the building. The magnitudes of the loads were reported previously in **Tables 6** and 7. The load effects created by the external wind forces were used to specify the hold-downs and shear connectors necessary to maintain continuity of MWFRS load path. The garage was not analyzed, but the procedure would be the same. To simplify the analysis, the end zone loads for case A were applied on both ends to simplify the analysis. To maintain a uniform balanced load in case B (wind applied to the gable end), a weighted average of 0.69 kN/m^2 (14.4 lbf/ft^2) was taken for zones A and C and applied horizontally. An average of zones E and F that was calculated to be −0.95 kN/m^2 (−19.8 lbf/ft^2) was applied vertically to the windward side of the roof, and an average of zones G and H that was calculated to be 0.61 kN/m^2 (12.7 lbf/ft^2) was applied vertically on the leeward side of the roof.

Analysis showed that structural connectors were needed for the roof, but not for the floor-to-floor connections and the foundation connection. Connectors for the truss ends must be able to simultaneously transfer uplift and north-south shear loading as well as shear loading alone in the east-west direction. Simpson Strong Tie (SST) H2.5A hurricane connectors were considered for the truss end-to-top plate connection. This connection resists both shear and uplift. The H2.5A has a shear capacity of 0.58 kN (130 lbf) and uplift capacity of 1.62 kN (365 lbf). The truss end loads are, respectively, 0.18 kN (40 lbf) and 0.27 kN (60 lbf). Applying a unity equation, the demand/capacity ratio is 0.18 kN/0.58 kN + 0.27 kN/1.62 kN = 0.477 < 1.0; therefore, the connector is adequate. An example of a typical truss connector is shown in **Figure 12**. SST A21 angles were considered for the gable end truss-to-top plate connection. This connection is subject to a total shear load of 10.7 kN (2400 lbf) when the wind is applied perpendicular to the gable end. SST A21 has a design capacity of 1.09 kN (245 lbf) per connector; therefore, the required number of connectors will be 10.7 kN/1.09 kN, which gives approximately 10 connectors.

The structure was checked for overturning at the second floor and at the first floor. The weight of the structure was adequate to resist the overturning moment in both locations. Sliding was only checked on the roof to specify the structural connectors. Sliding on the second floor is resisted by the nailed connection between the bottom plate and the floor assembly. Typically, there are sufficient nails engaged to resist the shear force. As for the building-to-foundation connection, there is no reason to expect an extraordinary loading at this junction, so anchor bolts are specified according to IRC Chapter R403.1.6. The I-joist to soleplate toenail connection was not checked in this analysis but should be checked in an actual design.

Figure 12.
Typical truss-to-top plate structural connector.

4.16 Wall bracing

Wall bracing for residential construction typically involves designating sections along the exterior wall length as shear panels. Structural wood panels are used on the exterior side of the wood framing, and gypsum wallboard on the interior provides the shear resistance and load transfer capability. Plywood or OSB is typically used for the wood structural panels. IRC Table 602.3(3) prescribes a 9.5 mm (3/8 inch) minimum structural panel thickness for 406.4 mm (16 inch) O.C. stud spacing; however, the builder prefers a 11.1 mm (7/16-inch)-thick OSB panel, which is required to be fastened to framing using 8D common nails at 152.4 (6 inch) O.C. around the perimeter and 304.8 mm (12 inch) O.C. in the field.

IRC Section R602.10 will be used to specify shear panel length and location along the wall line. Section R602.10 has provisions for various wall bracing methods. The bracing in this home will follow the requirements for the intermittent wood structural panel (WSP) method or one of the continuous sheathing methods. Because this home is categorized in seismic design category A, Section 602.10.1 allows for different methods to be used along different wall lines. Different intermittent methods could even be used along the same wall line in this category, but if using any of the continuous sheathing methods, the whole wall line must be continuously sheathed.

For the design of this home, it was more economical to use the WSP method for the majority of the shear panels. Section R602.10 requires 609.6 mm (24 inch) corner returns or braced panels at the end of each wall. At least one of the corners does not meet this criterion. When this occurs and the designer is using the continuously sheathed wood structural panel (CS-WSP) method, Section 602.10.4.4 requires the use of 3.56 kN (800 lbf) hold-down devices in lieu of a 2 foot corner return. This is often costlier than the extra amount of sheathing required for the WSP method. Another issue to consider when specifying wall bracing is the stud spacing. In this home, the studs are spaced at 406.4 mm (16 inch O.C.); therefore, it is prudent to specify shear panels 406.4 mm (16 inch) increments, even though the requirements may be less. The location of the shear panels is specified in the drawing set located in Appendix A.

4.17 Horizontal floor diaphragms

The floor assembly is treated as a flexible diaphragm when transferring lateral loading. Wind is transferred from a tributary area of the exterior wall to the rim board of the floor assembly and then into the structural sheathing. The floor sheathing then transfers that load to the exterior shear walls (structural panels within the wall system) parallel to the wind direction below the floor assembly. The diaphragm is treated like a deep beam for the purposes of analysis. The reactions are the connections with wall below. The floor assembly deflects, which causes tension and compression forces called chord forces in the walls below, which are perpendicular to the wind loading. The sheathing layout and the attachment of the sheathing to the I-joists have the greatest effect on the strength of the diaphragm. In this case, the floor sheathing and the required nailing were specified from the IRC in the floor assembly section of this report.

4.18 Connections

Most connections in wood-framed homes are made up of nailed connections. The majority of the connections in a typical home can be found in IRC Table R602.3. The items specified from the IRC in this wood-framed section are based on compliance with this table. In this study, only a few of the typical critical connections for the structural system were specified.

5. Concluding remarks

This chapter presented a complete design of a typical US single-family home made of conventional wood-frame system. Initially, the applicable building and material codes were introduced and relevant provisions discussed. A typical home plan by a PA builder was discussed and explained for detailed design. The process of load selection and load path and load combination was discussed. Then based on application of the resultant loads on typical structural elements, detailed designs for roof sheathing, roof trusses, exterior walls, main wind force resisting system, floor system, girders, columns, and foundation walls and footings were presented and discussed. Where appropriate, tips and guidelines for typical design were offered so that the procedure presented can be followed by designers as appropriate. While other structural systems are becoming increasingly available, the wood-frame system is still the dominating system as in the USA lumber is readily available at highly competitive process. This makes structural systems other than conventional wood-frame less competitive, unless there are special conditions where cost may not be the main determining factor.

Appendix: design drawings

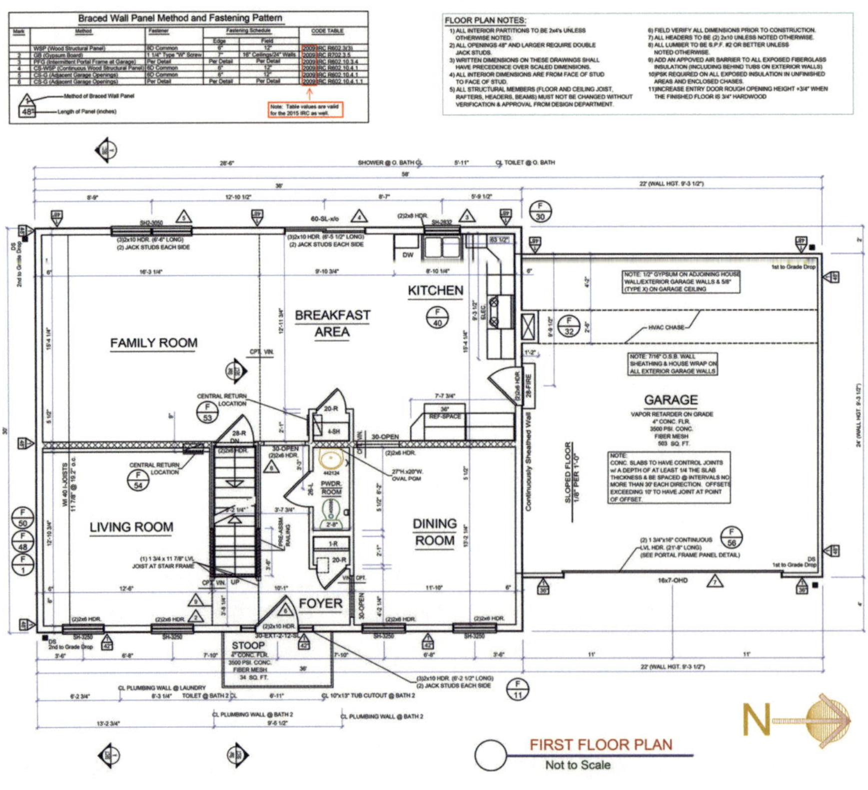

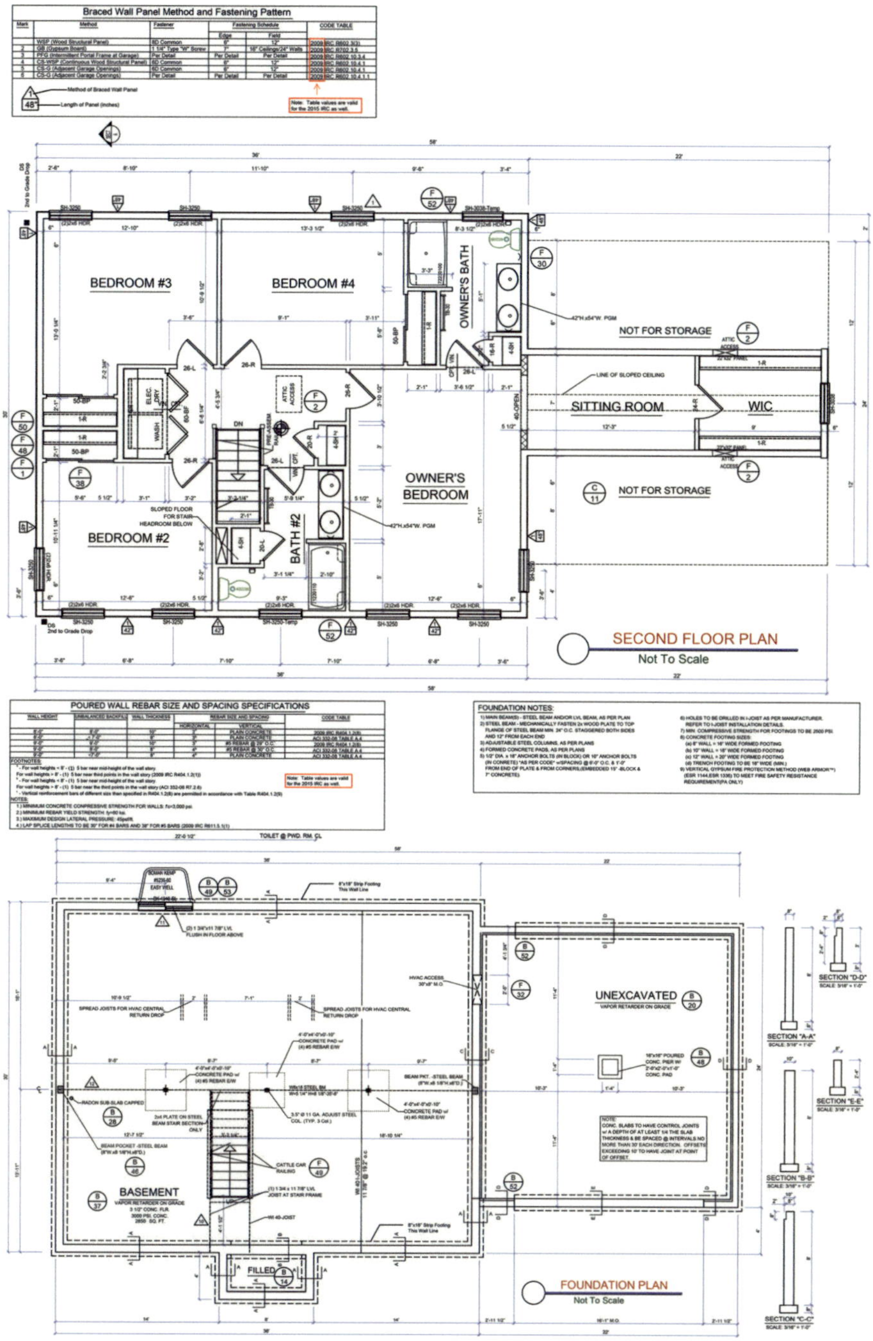
Braced Wall Panel Method and Fastening Pattern
BEDROOM #3
BEDROOM #4
OWNER'S BATH
NOT FOR STORAGE
SITTING ROOM
WIC
OWNER'S BEDROOM
BEDROOM #2
BATH #2
NOT FOR STORAGE
SECOND FLOOR PLAN
Not To Scale
POURED WALL REBAR SIZE AND SPACING SPECIFICATIONS
FOUNDATION NOTES:
UNEXCAVATED
BASEMENT
FILLED
FOUNDATION PLAN
Not To Scale

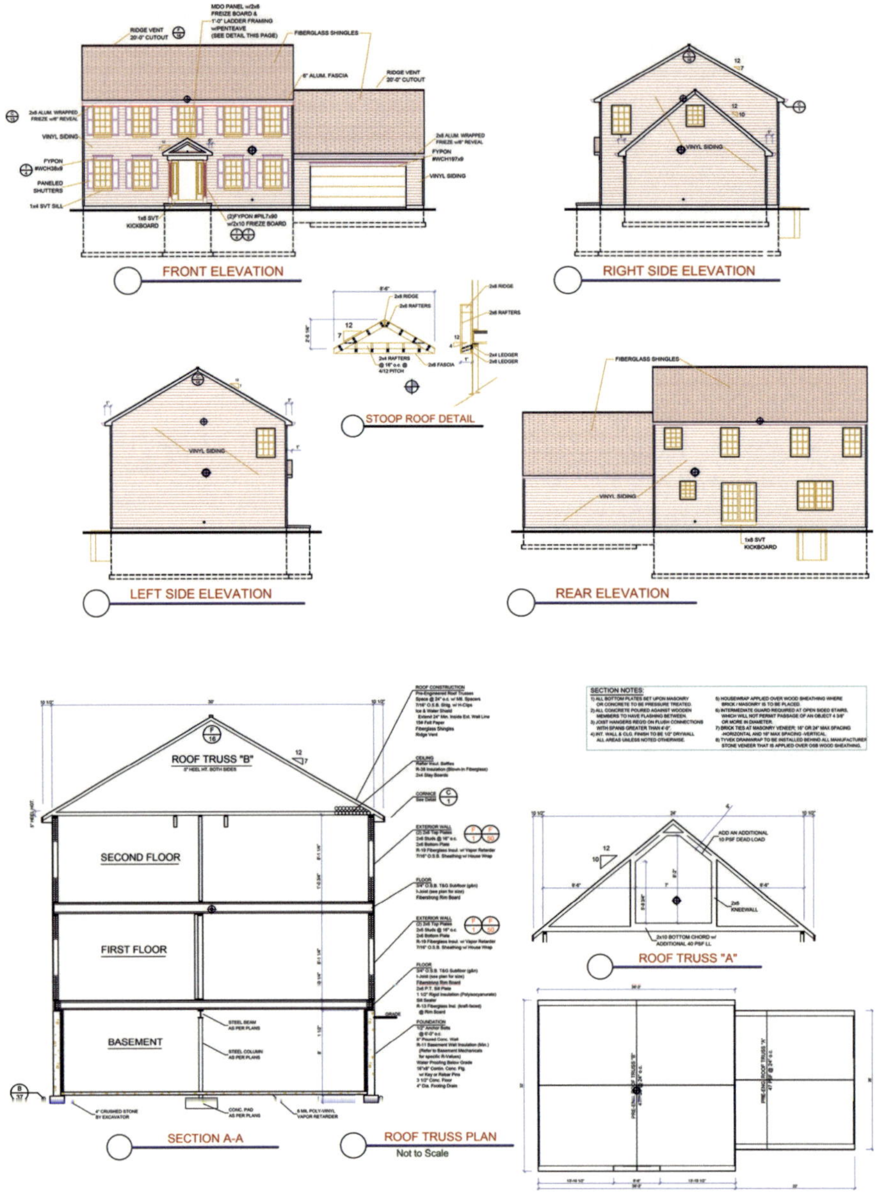
FRONT ELEVATION
RIGHT SIDE ELEVATION
STOOP ROOF DETAIL
LEFT SIDE ELEVATION
REAR ELEVATION
ROOF TRUSS "B"
SECOND FLOOR
FIRST FLOOR
BASEMENT
SECTION A-A
ROOF TRUSS PLAN
Not to Scale
ROOF TRUSS "A"
SECTION NOTES:
FIBERGLASS SHINGLES
VINYL SIDING

Author details

Anthony C. Jellen[1] and Ali M. Memari[2,3]*

1 Jellen Engineering Services, Mechanicsburg, PA, USA

2 Department of Architectural Engineering, Penn State University, University Park, PA, USA

3 Department of Civil and Environmental Engineering, Penn State University, University Park, PA, USA

*Address all correspondence to: memari@engr.psu.edu

References

[1] ICC (International Code Council, Inc.). In: 2015 International Residential Code for One- and Two-Family Dwellings; Country Club Hills, IL. 2014

[2] Rutkowski H. ACCA (Air Conditioning Contractors of America). Manual J Residential Load Calculation. 8th ed. Arlington, VA: Air Conditioning Contractors of America; 2015

[3] ICC (International Code Coucil, Inc.). In: 2015 International Building Code; Country Club Hills, IL. 2014

[4] ASCE (American Society of Civil Engineers). Minimum Design Loads for Buildings and Other Structures. Reston, VA: ASCE 7-10; 2010

[5] AWC (American Wood Council). Wood Frame Construction Manual (WFCM) for One- and Two-Family Dwellings 2015 Edition. ANCI/AWC WFCM-2015; Leesburg, VA. 2014

[6] NAHB Research Center. Inc. HUD (U.S. Department of Housing and Urban Development). Residential Structural Design Guide. Washington, D.C.: PD&R - U.S. Department of Housing and Urban Development; 2000

[7] ICC (International Code Coucil, Inc.). 2015 International Energy Conservation Code; Country Club Hills, IL: International Code Council, Inc.; 2014

[8] Breyer DE, Fridley KJ, Pollock DG Jr. Design of Wood Structures—ASD. New York, NY: McGraw-Hill; 2003

[9] Weyerhaeuser. Trus Joist TJI 110, TJI210, TJI230, TJI360, TJI560 and TJI560D Joists Specifier's Guide,#TJ-4000; Seattle, WA. 2018

[10] Mitek Industries, Inc. USP Product Catalog 59th Edition,#2515; Minneapolis, MN. 2018

[11] Simpson Strong-Tie, Inc. Wood Construction Connectors 2019-2020, C-C-2019; Pleasanton, CA. 2019

[12] AISC (American Institute of Steel Construction). Steel Construction Manual. 14th ed. Chicago, IL: American Institute of Steel Construction; 2011

[13] ACI (American Concrete Institute) Committee 332. Code Requirements for Requirements for Residential Concrete and Commentary. ACI 332-08; Farmington Hills, MI. 2008

[14] ACI (American Concrete Institute) Committee 318. Building Code Requirements for Structural Concrete and Commentary. ACI 318-14; Farmington Hills, MI. 2014

[15] Lindeburg MR. Civil Engineering Reference Manual. 11th ed. Belmont, CA: Professional Publications, Inc; 2008

Chapter 4

Communicating Added Value in Wooden Multistorey Construction

Cecilia Mark-Herbert, Elin Kvennefeldt and Anders Roos

Abstract

Climate changes point to the needs to find sustainable materials for residential multistorey housing as a growing proportion of populations across the world live in urban areas. Despite positive environmental effects, wood has a limited use in multistorey constructions even in countries with a strong tradition to use wood in residential housing, such as Sweden. As new materials, techniques were developed and studies of properties of wood as a construction material were communicated, and legislation was altered in Sweden in the mid-1990s, allowing for the use of wood in multistorey housing. The expected market growth was slow and uneven even when incentivizing programs were developed. This chapter explains consumer perspectives in a town, Växjö, where the tradition of using wood in multistorey construction is strong. It points to the needs of knowing more about consumers' perspectives—in order to communicate added values, that is, environmental benefits, in suitable market channels.

Keywords: consumer perceptions, product-dominant logic, residential construction, service-dominant logic, sustainable development

1. Sustainable development in residential construction

The world population is projected to increase by more than 1 billion people in the next 15 years, reaching 8.5 billion in 2030 and 9.7 billion in 2050 [1]. This forecasted growth in population points to an increase in demand for housing where a growing proportion of the population will live in cities. The projection points to an increase from the current 54 to 66% living in cities by the middle of this century [2]. Cities will continue to be the arena for residence and for commercial activities, a place with needs for sustainable changes.

Sustainable construction, for residential and commercial needs, can be seen as an opportunity for the construction industry. According to Bordeau [3, p. 364], the main challenge is "*to transform the demand for sustainable development into an opportunity, to create and access new markets, and to innovate responses which satisfy traditional industry demands and the new societal demands for sustainable development.*" It is claimed that the construction and utilization of buildings could be reduced as much as 40% of the energy consumption, the use of extracted materials can be reduced by 50%, and the total GHG emission could be reduced 35% [4]). The call for sustainable methods in construction is clear, and goals are set up globally in Sustainable Development Goals as well as locally in political goals and corporate objectives.

In recent years, a positive spread of wooden multistorey construction (**WMC**) in the Nordic countries points to business opportunities as part of an emerging forest-based bioeconomy [5]. However, changes in construction practices in multistorey construction for residential use are slow. The aim of this chapter is to explain factors that influence consumer behavior, focused on purchasing an apartment in a wooden multistorey construction (house), which ultimately may decide how markets for WMC develop.

The following text starts with a brief overview of factors that influence the development of markets for wood-based residential construction. Perspectives on wood as a construction material, policy aspects, and market development serve as a background for understanding how consumer perceptions are developed. The following subchapters account for a theoretical understanding of consumer behavior, an empirical study in light of local market development and a discussion of what can be learned from the case study with regard to the consumers' roles in the development of markets for WMC. A case study of Växjö in Sweden serves as inspiration for learning about market development for WMC.

1.1 Construction materials

Perceptions of physical properties of wood-based materials influence the current practices as well as the development of policies that regulate the use of wood in construction. Properties of wood can be divided in accordance with a number of factors related to, for example, the production of the raw materials (wood), use of raw materials in construction of houses, use of the house or the apartment in the house itself, and last but not least, deconstruction of the house when it is no longer needed. A brief overview of wood properties that have a bearing on its use in construction from a product-dominant logic perspective is provided in **Table 1**.

The list of characteristics of wood can be made much longer, but these properties have a bearing for the empirical study and the context in which it is conducted (Sweden). The perceived characteristics are by no means static, which indicate that as new technologies, products as well as techniques, are developed, some of the perceptions of wood are renegotiated, which may lead to changes in policy and legislation.

1.2 Markets

Historically, wood and stone have served mankind as construction materials for residential housing as well as commercial buildings. The tradition of building residential single houses in wood is well established in Sweden [6] where about 90% of new residential houses are built in wood (wooden frames and/or wood exterior and interior as a major part of the construction material). However, the development of multistorey constructions has been slow in progress in spite of promotional activities to spur market development.

Since 1994 when the legislation for multistorey construction was altered in Sweden, to allow for wood as a construction material (again) in multistorey housing in Sweden, the market has been very slow in development. Promotional activities sanctioned by the Ministry of Industry in 2002 were followed up by more local initiatives, neither of which led to any major market developments. Lately, however, additional governmental efforts, based on political objectives in line with Sustainable Development Goals (**SDG**) and an increased insight in how forestry may serve in the development of bioeconomies, are made to promote wood as a construction material for WMC, for local as well as international markets [7].

Characteristics	Effects	Perspective
Renewable, part of a circular bioeconomy	Carbon sink, an arena for ecosystem services (other productions on the same land), part of bioeconomy system and partial solution to some of the Sustainable Development Goals (SDGs) in more local political objectives and agendas	Society
Traditional and suitable material for house construction	Traditional material that is culturally grounded and part of architectural practices Skills are well developed for family housing construction but less so for multistorey house construction Part of the local construction culture Suitable (light weight) material for prefabrication and fast on-site assembly	Society Industry Consumer
Physical properties with regard to temperature, sound, and weight	Product and process innovation has paved the way for fire-resistant alternatives and safe use of wood in construction frames Offers possibilities for modular prefabrication construction where the efficiency in material use is higher Flexibility in the construction process makes it easier to adapt to the physical context for the construction. Wood is also a lighter material than concrete, which points to usefulness when the grounds are not stable Wood materials offer a nice atmosphere when used internally in a construction. Moisture, fire, and sound properties need to be taken into account in the adaptation of the production as well as use of the house	Society Industry Consumer
Availability and ownership of wood materials	Wood is naturally occurring in Sweden and the ownership structure promotes development of markets. In the future, forest-based resources may be given additional importance as part of creating bio-based economies	Society Industry Forest owner

Table 1.
Perceptions of wood properties in light of use in multistorey construction based on a review with minor modifications from Sjöström [6].

With a forecasted annual need for 93,000 new homes per year in Sweden, the predicted construction of 120,000 homes for 2018–2019 does not cover the forecasted market needs. Of these new homes, about 10% are currently flats in WMC apartment houses [6]. With an understanding of wood as a construction material and a political will to develop WMC markets, what are the views of consumers?

1.3 Needs to learn about consumer perceptions

Consumers' views are greatly influenced by norms in society [8]. An increased awareness of human influence on sustainable development is reflected in societal dialog in channels such as media, social media, NGO influences, and more traditional elements of culture such as family and traditions. Consumer expectations, values, and communicational traditions are therefore of tremendous importance for understanding decision-making in purchasing an apartment (often referred to as an example of an investment product).

Communication is seen as key in the purchasing process since it concerns a lot of money, and a product that consumers may not be capable of evaluating all aspects of *ex ante*. We assume that consumers need information to make a well-grounded decision about a major investment, their new home [9]. Given the understanding of wood properties, a need for housing, and consumers' need for information about an investment, their home, our aim is to explain the perceived advantages among apartment owners in wooden multistorey houses.

Themes for questions	Theoretical starting point
Awareness of living in a WMC apartment Where did you get information about the wooden construction of the house?	*Consumer awareness of alternatives in consumption* [10]
Information about climate-related properties of a wooden house	*Information about wood properties* [11, 12]
Channel for information (about wood properties and climate properties)	*Channel for communication* [8]
Factors that influenced the decision to buy the apartment	*Decision-making* [10, 13, 14]

Table 2.
Themes for the interview with residents in apartments in multistorey constructions, in Växjö and theoretical origin for the theme [9].

1.4 Approach

An empirical case study was conducted with the ambitions to investigate consumer perceptions of the apartment they own (in a wooden multistorey house) [9]. A small town in Sweden, Växjö, was selected because the WMC tradition was well developed. Växjö is often referred to as "wood construction town", and it is seen as a forerunner for urban development in Sweden.

Three WMCs were selected (Strandsnäckan, Wälludden, and Portvakten), and in these houses, (11, 14, and 17, respectively), randomly selected residents were willing to contribute to the project in an interview (in February–March, 2018). After receiving an informed consent from the respondents and a promise for the respondents to be anonymous, the interview was made as a leisure dialog, based on open-ended thematic questions. Themes for the interview covered were connected to factors that the literature review had pointed to as key for making a residential purchasing decision (**Table 2**).

The result from the interviews in the selected houses and randomly selected residents does not lend itself to generalization. It should be seen as indicative of what might have explanatory power and what we need to investigate further.

2. Understanding consumer perspectives

Market development can be explained from various perspectives. Policy instruments, such as taxes, information campaigns, and legislation play instrumental roles in promoting desired behaviors. Industrial development such as research and public private partnerships may also account for market development. In this case, our focus rests on the role of the consumer.

2.1 Marketing

A traditional view of a theory for consumers' purchasing behaviors for investment products is described in these phases [10] as part of a planned behavior (theory of planned behavior):

I. Need recognition

II. Searching and gathering information

III. Evaluating the alternatives

IV. Actual purchase of the product or the service

V. Postpurchase evaluation

Although the process appears to be linear, it is normally an iterative process, in which the consumer gains additional understanding of choices made and choices to be made. Communication fills important roles throughout the purchasing process [13]. It may seem strange that we did not focus on phase II–IV but rather on V, the after-purchase evaluation in our study. This is explained by our understanding that consumers' postpurchase behavior is assumed to influence their future purchases as well as in their roles as ambassadors, in communication with other consumers.

2.2 Marketing communication

The marketing mix (product, price, promotion, and place) [10] serves as a starting point in the purchasing process. This model offers an understanding of marketing from a producer perspective. The classical model is reinterpreted, focusing on the four P's in terms of customer perspectives as *customer solutions, customer costs, communications, and convenience* [13] in order to understand consumer views. The big difference in perspective, from P:s to C:s, deserves to be emphasized as it gives the consumer a key role in a possible market development for the WMC apartments.

3. The housing situation and markets for multistorey construction (WMC) in Sweden

The markets for residential constructions, houses as well as apartments in multistorey constructions in Sweden, are changing slowly and it does not meet captured in the word, *shortage*. The housing situation varies from one municipality to another, which in part can be explained by political ambitions and by corporate investments. In the metropolitan regions of Stockholm, Gothenburg, and Malmö, housing construction is expected to increase significantly over the coming years to meet the current housing shortage in the country [15] illustrated in **Figure 1**.

The needs for residential house construction in **Figure 1** are clear. According to this forecast, some 90,000 additional homes are needed annually between 2018 and 2024 [15, p. 20]. Looking more closely on what has been completed in terms of residential houses since the late 1930s (**Figure 2**) further supports the understanding of market development needs [16].

Figure 2 shows the number of apartments and single-family houses since the 1940s in Sweden. The new residential construction has been below the needs of markets since the 1970s. The graph also indicates that since 2010 the rate of construction for multistorey houses (apartment homes) is increasing. In 2017, some 51,500 new homes were completed in Sweden, which is a much needed increase. This is explained by a population growth and continued movement of individuals to urban contexts, especially metropolitan areas, from rural areas. However, even with this increase in new homes, it does not cover the forecasted needs of 90,000 new (additional) homes on an annual basis [15].

3.1 Traditions of using wood in construction

A tradition to use wood in house construction in Sweden is well established. Of the approximately 10,000 new single houses that are built in Sweden on an annual basis, some 90% are built in wood [17]. The situation for multistorey construction

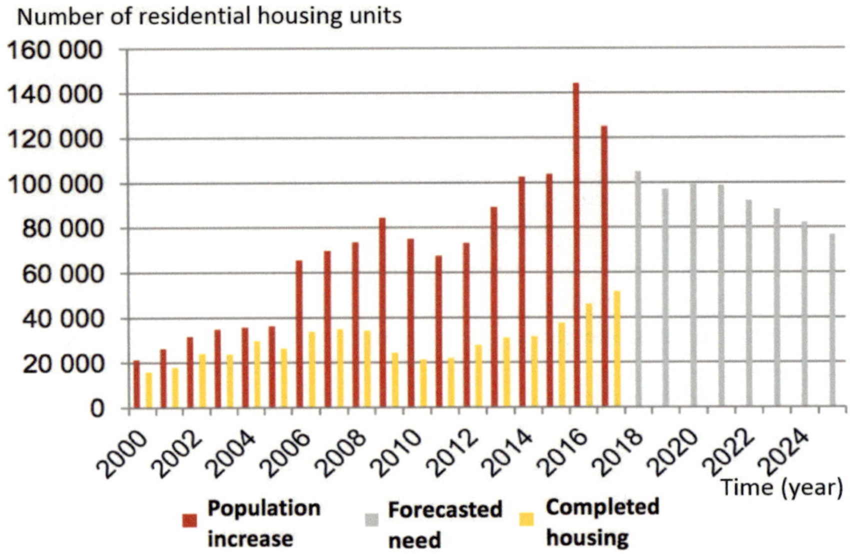

Figure 1.
The relationship between completed housing construction and the forecasted need for housing up to the year 2025 [15].

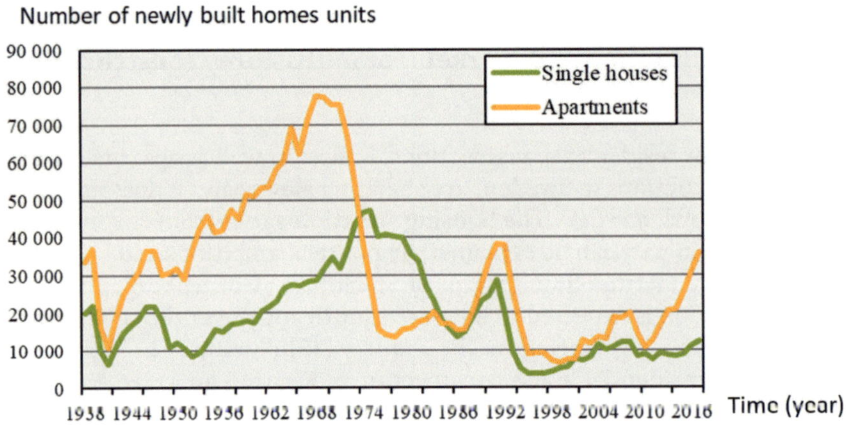

Figure 2.
Number of completed apartments in multifamily houses and single-family houses from 1938 to 2017 in Sweden [16].

is quite different. Due to fear of fire, a legislation has been limiting construction in wood (over two levels). However, since 1994, it is permitted to use wood as a bearing material in multistorey residential construction in Sweden. Development of new techniques, materials, and tests of the properties of wood in construction has proven that wood serves as a safe and esthetically pleasing material, in addition to the properties that benefit sustainable development.

The ban to use wood in multistorey construction has had consequences on the market in terms of a deprived skills and technological development [13]. Other construction materials, such as concrete and steel, have taken the lead in practices for multistorey construction. When the legislation was altered in 1994, the market did not respond, which is explained by a change of practices and skills in major stakeholder groups (construction workers, engineers, architects as well as in procurement stakeholders representing municipalities) [18].

Political efforts were made to support the reestablishment of wood practices in multistorey construction in Sweden. In 2002, the Ministry of Enterprise and Innovation presented a proposal to develop a strategy for increased wood construction in Sweden, "More wood in construction DS2004:1" [19]. It was based on the understanding of a need to meet needs of needs of homes located in urban areas, land use, and sustainable development objectives for construction. In spite of political efforts on national and local levels, the market development has remained very moderate.

In 2018, the Swedish Ministry of Industry presented a policy document "Focus on Wood Construction" [20], which clearly voices the understanding of needs for multistakeholder dialogues in order to promote a market development. It points to key stakeholders as forest-based industry corporations, municipalities, architects, and consumers as drivers of a change in construction practices. Representing the government, representatives of the Ministry describe perceived benefits in terms of creating new jobs, technological development, and increased housing production, with an understanding of these outcomes as part of sustainable development.

At a local level, Swedish municipalities address political objectives related to sustainable development and residential construction with different objectives. One of the municipalities that has worked with wood in multistorey construction in a very structured way is Växjö.

3.2 Consumer perspectives in Växjö

The town, Växsjö, is often referred to as a "wood construction town" with an ambition to be a center for wood construction expertise. It is located in the southern part of Sweden, surrounded by forests, which explains the well-developed industrial system that uses the forest resources. Municipal ambitions of sustainable development efforts as well as traditions may account for politically well-founded public procurement efforts for wooden multistorey constructions (WMCs). Their local strategies have supported WMC for a long time [21] with ambitious goals set for 50% of the new constructions of multistorey homes to be in wood by 2020. Municipal bold objects set in 2013 and efforts to follow up on progress have resulted in 44% new WMC already by 2015 [22]. The efforts continue to reach the 50% target by 2020 [23].

Assuming that municipal politics and actions reflect consumer values and needs, this pilot study set out to investigate consumer perspectives of WMC. The interviewed residents of apartments in WMC houses were interviewed, and **Table 3** provides an overview of the results [9]. The 42 interviews in three residential areas are merged, as they did not show any major differences from one residential area to another.

Empirical results (**Table 3**) from the interviews indicate that a majority of the apartment residents are aware of that they live in a WMC apartment. The source of that information is less clear, which might be explained by some time having passed since the purchase was made or not being the first owners of the apartment.

Channels for information about climate properties of wood construction is preferred in word of mouth dialogs, web pages, printed matters, and media information in general.

The last theme for the interview, the factors that were of major importance for purchasing an apartment or not, points to the importance of the location of the apartment. Size, price, and atmosphere also play important roles. The last listed factor, environmental properties, offers some food for thought.

Thematic questions	Results from interviews (in total 42)
Awareness of living in a WMC apartment Source of information about WMC apartment	34 knew they lived in a WMC apartment 6 thought it was a concrete house 3 did not know about the construction material 12 persons are aware of having been given information about WMC Oral information from sales agent and printed information from the house association
Information about climate-related properties of a wooden house	14 are aware of having been given information 22 do not recall any information 6 do not know
Channel for information (about wood properties and climate properties)	21 prefer to get information in personal dialogs (word of mouth) 12 would prefer to get information on a web page 11 request printed information and 7 reply that TV and radio is a good channel for communication
Factors that influenced the decision to buy the apartment	The location of the apartment (34) Size of apartment (16) Price (13) Atmosphere (13) Environmental properties (6)

Table 3.
Results from interview with 42 residents in apartments in wooden multistorey constructions (WMCs) in Växjö (February–March, 2018) [9].

4. Communication and market development

The development of a market can be seen from many perspectives. Starting with a product-dominant logic, focusing on the product (an apartment in a WMC house), we continue the discussion with a more service-dominant logic [21] perspective focusing on societal and consumer needs.

4.1 A product-dominant perspective—consumer behavior

This project set out with an understanding of the importance of commutation as means to promote consumer behavior. Previous research on consumer awareness in purchasing apartments in WMC [24–29] points to shortcomings in communication regarding climate effects.

Our empirical study shows that efforts to communicate have been made, which is to be expected in a "wood town" like Växjö, but the consumers' recollection of the information from the purchasing situation was not very clear. This may be explained by some time having passed since the purchasing decision was made, information materials that were not tailored to the needs of the consumers, the interviewee not being the person that made the purchasing decision or information that was provided in a communication channel that was not suitable. The level of awareness of benefits related to wood as a construction material in WMC appears to be limited. Consumers' limited insight to technical production-related benefits is expected, but the awareness of physical properties in use and environmental benefits with WMC also appears to be moderate among the interviewees. There is, clearly, room for improvements if consumer awareness is seen as important.

The last question in the empirical study, about factors that determine a purchasing decision of an apartment in a WMC house, points to needs to rethink our understanding of what influences market development. The first thing we need to problematize is that of the difference between *preference* and actual *purchasing decision*. Housing preferences are influenced by a number of factors [14], for example, household size, transparency of the housing market, availability of alternative housing options, and lifestyle-related factors. All of the above, in addition to personal budget constraints, financial institute conditions for taking loans and governmental policy systems that may or may not offer tax deductions and subsidies, on the other hand, influence purchasing decisions. It seems that studies of conditions of consumer awareness of environmental aspects of WMC require a contextual understanding of institutional conditions for making decisions.

Worth noticing is that the consumers did not mention health aspects influencing their purchasing decision in our empirical study, which is somewhat surprising. After all, the indoor environment is important for an individual's life, especially in temperate climate such as that of Sweden. Burnard and Kutnar [28] point to health aspects in terms of wood as a construction material that may relax stress.

Given the current shortage of housing alternatives in Sweden [15], there might not be a need to communicate WMC benefits. Consumers in need of an apartment will buy it, independent of construction materials used in the house [9]. Furthermore, the importance of location points to that this will be key if consumers are given alternatives.

As pointed out by Jansen et al. [14], availability of housing alternatives plays a key role. If there are no WMC apartment available in a particular geographical area, of a requested size and price range, that will limit the consumer's possible choice of a WMC apartment. The logic is the same as that of buying food in a food store. It is the management's portfolio decisions in a food store that will set the limits to what products that I may purchase in that food store.

4.2 A service-dominant perspective—societal and consumer needs

In this case, goals for expanding markets for WMC were set on a national level as a way to meet forecasted housing demands, preferences, and work toward sustainability objectives in an industry that uses a substantial amount of resources. Some municipalities, like Växjö, picked up on these national objectives and made it their local agenda, but the vast majority of municipalities in Sweden have not.

A service-dominant logic starts with consumer needs and looks for ways in which these needs can be met. With an understanding of consumers' limited capacity to influence the material choice in house construction, it seems that other economic and macrolevel incentives are needed. Standards and sustainability ecolabels would be one way to go [30], but Hurmenkoski et al. [27] call for regulatory frameworks to support the WMC market developments. Although the environmental benefits in using wood for house constructions are associated with societal benefits [31], it cannot be assumed that consumers fully comprehend the benefits or are willing to pay extra, which points to needs for policy instruments to be used such as taxes and subsidies to promote market development in a direction toward sustainable development.

Looking into the future, Høibø et al. [32] suggest that future generations are more environmentally aware when it comes to paying attention to house construction materials. Their findings support other researchers' views [8, 13, 14] that younger generations, the future inhabitants of apartments in WMC houses, future generation of business leaders, and politicians will be more aware of SDGs in their private life as well as in their professional conduct. Given the lasting character of a house that may last for hundreds of years, it is important to include educated guesses of where the future is taking us in strategic choices.

5. Conclusions

Communicating climate effects of WMC appears to be critical in business to business (B2B) relations, in private and public procurement. Consumers are not making the portfolio decisions related to building WMC houses. It is the construction agents, municipal planning board, and suppliers in the construction industry, like architects, construction planners, and construction workers who set the market in which consumers usually only can make a choice of an apartment. That is to say that in a town that does not have political objectives to develop WMC, there will be less apartment in wooden houses (WMC) and the consumer would have to make a choice from what is available on the market.

In the presented case of Växjö [9], bold political objectives were set up and communicated. They serve as an action plan and a collective contract to work toward sustainable development in the construction of residential housing as part of what Beltz and Peattie [13] refer to as *sustainability marketing transformations.*

Shortcomings in development of a housing market with options for the consumer to make a choice from (WMC or concrete buildings) are clear. They can be explained by a number of factors, for example, relating to:

- current practices in the construction industry (ongoing contracts and habitual practices) [3]
- organizational arrangements [6]
- skills that are built up over time using the current materials [18]
- training programs of professionals, architects, construction planners as well as construction workers [18, 33]
- political objectives and priorities (what is built where) [3, 29]
- the current housing situation reflected on the market [14]

All of these factors may serve as market development constraints, and they need to be further investigated to provide an understanding of the context for WMC market developments.

In conclusion, wood offers superior qualities with regard to developments of circular bioeconomies, as part of sustainable development. Consumer awareness of WMC as means to make choices in support of sustainable development is important. Consumers may not influence the current market offers directly in their purchasing decisions, in a short time frame. They will have to make a choice among the current offers if they need a house or an apartment now, but they may still influence political policies and actions in voting procedures, NGO engagement, and taking the role as a WMC ambassador in everyday dialogs.

Continued research is needed to explain policy implications of stimulating the market for WMC. Examples of such areas have to do with organizational aspects as well as policy implications of market transitions toward sustainable development, where an increased use of WMC is seen in models for circular bioeconomy. Organizational aspects of markets where the WMC has been successful are associated with collaboration efforts in so-called public-private partnerships [6]. Experiences from recent political efforts of market transitions [13, 34] are seen in banning the old light bulbs, stimulating markets for solar panels and bicycles with batteries through subsidies. All of these examples might serve as inspiration to forecast how policy instruments can be used efficiently to promote sustainable development in the development of WMC markets.

Author details

Cecilia Mark-Herbert*, Elin Kvennefeldt and Anders Roos
Department of Forest Economy, Swedish University of Agricultural Sciences, Uppsala, Sweden

*Address all correspondence to: cecilia.mark-herbert@slu.se

References

[1] United Nations Department of Economic and Social Affairs, Population Division. World Population Prospects: The 2015 Revision, Key Findings and Advance Tables. Working Paper No. ESA/P/WP.241. 2015. Available from: https://esa.un.org/unpd/wpp/publications/files/key_findings_wpp_2015.pdf

[2] United Nations Human Settlements Programme (UN-Habitat). Urbanization and Development: Emerging Futures. World Cities Report 2016. ISBN: 978-921-133395-4. Available from: http://wcr.unhabitat.org/wp-content/uploads/sites/16/2016/05/WCR-%20Full-Report-2016.pdf

[3] Bourdeau L. Sustainable development and the future of construction: A comparison of visions from various countries. Building Research & Information. 1999;**27**(6):354-366. DOI: 0.1080/096132199369183

[4] Herczeg M, McKinnon D, Milios L, Bakas I, Klaassens E, Svatikova K, et al. Resource efficiency in the building sector. In: Final Report for DG Environment. ECORYS and Copenhagen Resource Institute; 2014. p. 128. Available from: http://ec.europa.eu/environment/eussd/pdf/Resource%20efficiency%20in%20the%20building%20sector.pdf

[5] Toppinen A, Röhr A, Pätäri S, Lähtinen K, Toivonen R. The future of wooden multistory construction in the forest bioeconomy—A Delphi study from Finland and Sweden. Journal of Forest Economics. 2018;**31**:3-10. DOI: 10.1016/j.jfe.2017.05.001

[6] Sjöström F. Sustainable urban development through public-private partnership—Collaboration for increased wood construction [master's thesis no. 196]. Sweden: Department of Forest Economics, Swedish University of Agricultural Sciences. ISSN 1654-1367; 2017. Available from: https://stud.epsilon.slu.se/12857/1/sjostrom_f_171114.pdf

[7] The Swedish Government. Inriktning för träbyggande. 2018. Available from: https://www.regeringen.se/49ee7f/contentassets/37f07802672c45078a20d3a375e82c25/20180626_inriktning-for-trabyggande.pdf

[8] Ottman J. The New Rules of Green Marketing—Strategies, Tools and Inspiration for Sustainable Branding. UK: Greenleaf Publishing Limited; 2011. ISBN 978-1-906093-44-0

[9] Kvennfedt E. Kommunikation av klimatfördelar med flervåningshus i trä [master's thesis]. Sweden: Department of Forest Economy. Swedish university of Agricultural Sciences; 2018. Available from: https://stud.epsilon.slu.se/8121/

[10] Kotler P, Armstrong G, Parment A. Principles of Marketing Swedish Edition. 13th ed. Harlow, England: Pearson Education; 2011. ISBN: 9780273735960

[11] Eriksson L, Gustavsson L, Hänninen R, Kallio M, Lyhykainen H, Pingoud K, et al. Climate change mitigation through increased wood use in the European construction sector—Towards an integrated modelling framework. European Journal of Forest Research. 2012;**131**(1):131-144. DOI: 10.1007/s10342-010-0463-3

[12] Nord T, Brege S. Värden för världen – Konsekvenser av ett ökat industriellt träbyggande. Linköping: Linköpings universitet; 2013. Available from: http://www.mynewsdesk.com/se/sveriges-trabyggnadskansli/documents/rapport-vaerden-foer-vaerlden-28094

[13] Belz F-M, Peattie K. Sustainability Marketing, a Global Perspective. 2nd ed.

Chichester, UK: John Wiley & Sons Ltd; 2012. ISBN: 978-1-119-96619-7

[14] Jansen S, Coolen H, Goetgeluk R. The Measurement and Analysis of Housing Preference and Choice. Dordrecht, Netherlands: Springer; 2011. ISBN 978-90-481-8894-9

[15] NBHBP, Boverket/National Board of Housing, Building and Planning. Behov av nya bostäder 2018-2025. Rapport 2018:24; 2018. ISBN: 978-91-7563-573-6. Available from: https://www.boverket.se/globalassets/publikationer/dokument/2018/behov-av-nya-bostader-2018_2025.pdf

[16] SCB, Statistiska Centralbyrån/Statistics Sweden. Antal färdigställda lägenheter i flerbostadshus respektive småhus. 2018. Available from: https://www.scb.se/hitta-statistik/statistik-efter-amne/boende-byggande-och-bebyggelse/bostadsbyggande-och-ombyggnad/nybyggnad-av-bostader/pong/tabell-och-diagram/antal-fardigstallda-lagenheter-i-flerbostadshus-resp.-smahus/

[17] IVA, Ingenjörsvetenskapsakademien & Sveriges Byggindustrier. Klimatpåverkan från byggprocessen – En rapport från IVA och Sveriges Byggindustrier. Kungl. Stockholm: Ingenjörsvetenskapsakademien; 2014. ISBN: 978-91-7082-883-6

[18] Thiger E, Woxblom L, Roos A. Empathic design for wood product innovation based on genuine customer needs–A test application on Swedish builders. Wood Material Science & Engineering. 2015;**12**(3):118-128. DOI: 10.1080/17480272.2015.1056226

[19] Ministry of Enterprise and Innovation. Mer trä i byggande – Underlag för en nationell strategi att främja användningen av trä i byggandet. Departementsserien, Ds 2004:1. Regeringskansliet. Näringsdepartementet. 2004. Available from: https://www.regeringen.se/49bbba/contentassets/622a4cddc02a4026a3bc3c4f5d5b94aa/mer-tra-i-byggandet---underlag-for-en-nationell-strategi-for-att-framja-tra-i-byggandet-ds-20041

[20] Ministry of Enterprise and Innovation, Government Offices of Sweden. Fact sheet. Swedish Government initiatives to increased wood construction. Item no: N2017.46. 2017. Available from: https://sharingsweden.se/wp-content/uploads/2016/06/swedish-government-initiatives-on-wood.pdf

[21] Vargo S, Lusch R. Service dominant logic 2025. International Journal of Research in Marketing. 2017;**34**:46-67. DOI: 10.1016/j.ijresmar.2016.11.001

[22] Växjö municipality, Växjö den moderna trästaden. Växjö kommun träbyggnadsstrategi. 2018. Available from: https://vaxjo.se/download/18.157e2afb15d3ac8d0adb456/1500031035431/Tr%C3%A4byggnadsstrategi.pdf

[23] Växjö. Trästaden. Hållbart byggande i Växjö. 2018. Aavailable from: https://www.vaxjo.se/sidor/trafik-och-stadsplanering/oversiktlig-planering/planeringsunderlag/trastaden.html

[24] Bergkvist S. Trähusindustrins marknadsföring av klimatfördelar med trä – en studie om kommunikationen beträffande träbyggandets klimatfördelar. Uppsala, Sweden: Sveriges Lantbruksuniversitet, Fakulteten för skogsvetenskap. Institutionen för skogens produkter; 2015. Available from: https://stud.epsilon.slu.se/8121/

[25] Hu Q , Dewancker B, Zhang T, Wongbumru T. Consumer attitudes towards timber frame houses in China. Procedia—Social and Behavioral Sciences. 2016;**216**:s.841-s.849. DOI: 10.1016/j.sbspro.2015.12.081. Available

from: https://core.ac.uk/download/pdf/82081962.pdf

[26] Nilsson V. Komponenttillverkning i byggbranschen – en marknadsundersökning om prefabricerade huskomponenter och byggelement i byggbranschen. Uppsala, Sweden: Sveriges Lantbruksuniversitet, Fakulteten för skogsvetenskap. Institutionen för skogens produkter; 2017. Available from: https://stud.epsilon.slu.se/10184/1/nilsson_v_170504.pdf

[27] Hurmenkoski E, Jonsson R, Nord T. Context, drivers and potential for wood-frame multi-story construction in Europe. Technological Forecasting and Social Change. 2015;**99**:181-196. DOI: 10.1016/j.techfore.2015.07.002

[28] Burnard M, Kutnar A. Wood and human stress in the built indoor environment: A review. Wood Science Technology. 2015;**49**:969-986. DOI: 10.1007/s00226-015-0747-3

[29] Wang L, Toppinen A, Juslin H. The use of wood in green building: A study of expert perspectives from the UK. Journal of Cleaner Production. 2014;**65**:350-361. DOI: 10.1016/j.jclepro.2013.08.023

[30] Hansmann R, Koellner T, Scholz R. Influence of consumers' socioecological and economic orientations on preferences for wood products with sustainability labels. Forest Policy and Economics. 2006;**8**(3):239-250. DOI: 10.1016/j.forpol.2004.06.005

[31] Toivonen R. Product quality and value from consumer perspective – An application to wooden products. Jounal of Forest Economics. 2012;**18**(2): 157-173. DOI: 10.1016/j.jfe.2011.12.004

[32] Høibø O, Hansen E, Nybakk E. Building material preferences with a focus on wood in urban housing: Durability and environmental impacts. Canadian Journal of Forest Research. 2015;**45**(11):1617-1627. DOI: 10.1139/cjfr-2015-0123

[33] Hemström K, Mahapatra K, Gustavsson L. Perceptions, attitudes and interest of Swedish architects towards the use of wood frames in multi-storey buildings. Resources, Conservation and Recycling. 2011;**55**(11):1013-1021. DOI: 10.1016/j.resconrec.2011.05.012

[34] Gottberg A, Morris J, Pollard S, Mark-Herbert C, Cook M. Producer responsibility, waste minimisation and the WEEE directive: Case studies in eco-design from the European lighting sector. Science of the Total Environment. 2006;**359**:38-56. DOI: 10.1016/j.scitotenv.2005.07.001

Chapter 5

Operative Machinery Costs Analysis within Forest Management Implementation Frame

Francesco Carbone and Rodolfo Picchio

Abstract

Forest management affecting most of the world's forests is based on the implementation of forestry interventions. Their execution requires preliminary preparation based on technical documents, submission of the administrative procedures, and the execution of the forest harvesting. Market value of wood is achieved by determining the most probable transformation value. It is obtained as the difference of the revenues derived from the sale of marketable timber net of all the costs involved in transforming the tree into salable products. The chapter provides a theoretical framework of the evaluation approaches and the calculation methods of the timber value, considering the different forms of assignment of the forestry intervention to the logging company, as well as the types of ground and the types of product achievable based on the level of mechanization.

Keywords: legal aspects, transformation process steps, types of costs, stumpage value, entrusting approach

1. Introduction

Forests cover about 4.0 billion hectares in the world. All decisions and/or actions adopted in order to preserve, to conserve, and to harvest forests and trees within them Bettinger et al. [1] Grebner et al. [2] are expression of the forest management. It is also defined as a tool that forest-owners use to achieve social, economic, and environmental targets or also to implement sustainable forest management. Two of the following approaches characterize forest management:

- Monitoring approach, through observation, monitoring, elaboration data, and e-reporting of forest ecosystem state

- Technologic approach, through machines, tools, materials, and forest workers

The first approach concerns primary forest area, while the second is adopted in modified forests, planted forests, and other areas with trees that consist of 2.7 billion hectares (**Figure 1**) of the global forest area. For the modified forests, one of the most relevant actions is stand management. This management is

IntechOpen

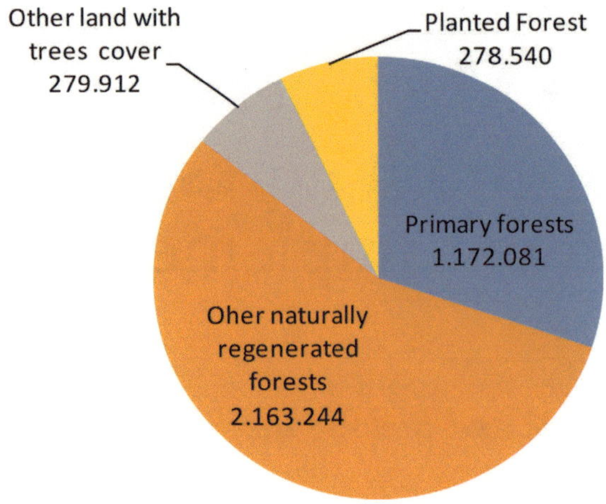

Figure 1.
The 4.0 billion hectares of forest area in the world articulated for anthropization categories (data × 1000 hectares). Source: FAO (2015).

- necessary, given the alteration (structural and compositional) these ecosystems have recorded over the centuries due to human activities;

- functional, to ensure the ecosystem perpetuity and guarantee overtime forest ecosystem services that influence the well-being of the humankind; and

- appropriate, in order to increase the resilient capacity of ecosystems currently under strong pressure due to socio-economic activities, climate change, and other global and local disturbance processes.

This action satisfies the aims of both forest landowners (FLOs) and logging companies (LCs). Using silvicultural criteria and proper manners, FLOs quantify the intervention in order to ensure the perpetuity of ecosystem (long-term vision) given by the forest's natural renovation capacity. LCs' point of view concerns the wood fraction exploitable (stand removal), which is the main output in timber transformation chains (short-term view). In the sustainable economic development policy, forests assume relevant roles [3]. That can be shortly explained as follows:

- Wood is a renewable natural resource.

- Timber is the main tangible output compared to other nontimber products.

- Timber in itself is an ecosystem service provider that provides bio-based resources and also biomass for clean energy production.

- Forest ecosystems guarantee regulative ecosystem services related to climate change contrast strategies and other global degradation processes.

- Forest ecosystems are areas where individuals or groups can live important social and sensorial experiences and increase their knowledge on natural life processes.

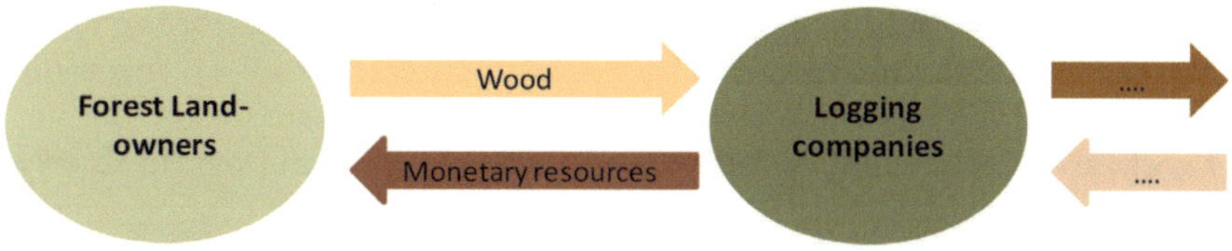

Figure 2.
Relationship between forest land owners and logging companies. Source: Our elaboration.

Using the Italian forest system as background, this topic has been developed in order to provide an international dimension.

A complementary relation exists between FLOs and LCs (**Figure 2**). FLOs are responsible overtime for the stand management. Silvicultural managements implemented during forest lifetime have influence on timber and nontimber products. However, FLOs normally do not have resources and knowledge to implement silvicultural management by themselves and sell timber products. The high investments needed for achieving an efficient and technological mechanization level, from an economic and financial point of view, would not be justified if the FLOs manage periodically small forest area, as the majority of FLOs in Europe. According to these evidences, FLOs entrust this job to specialized units, such as LCs, which have machines, technologies, materials, and workers with the knowledge of timber transformation process. Finally, LCs have proper knowledge to achieve the highest market price, given the market situation and the current economic trends.

Wood	It is the hard, fibrous, and structural tissues, composed of chains of cellulose, which forms the main substance of the trunk, branches, and roots of trees or shrubs.
Timber	It is the term to identify products obtained from the trunk, branches, and roofs of trees, at any stage after the tree has been felled. It includes the raw material, also known as rough timber or the processed material, used for construction, as firewood, and as other bio-based products.
Price	It is the amount of money that a buyer and a seller agree at the end of a negotiation to exchange between goods for money. In particular, there are a buyer willingness to pay and a seller willingness to accept in order to exchange the property right of goods and services. That information is collected specifically from the market.
Value	It is an amount (or datum) expressed in monetary metric, produced by experts or single individuals, using simple or complex elaborations, with the support of higher or lower technologies. Given a market price if it was subjected to any elaboration, example timber market price (€) multiply by quality (ton), the result is a value [€ × ton].
Forest management	It is all decisions and/or actions adopted in order to preserve, conserve, and use forests and trees within them.
Stand management	It is all decisions and/or actions related to the stand.
Silviculture management	It is the mode by which forest management is implemented.
Forest utilization	It is the action usually developed from the LC. Selected trees are felled and processed to obtain marketable timber.

Table 1.
Glossary of the terms used in the chapter.

Regardless of how stand management is entrusted to LCs, one of the mandatory steps is to determine the related costs. In dedicated literature [4–6], this topic is solved considering the operative cost only. However, differences exist between the type of costs and the calculation method, while transaction and overhead costs and revenues are not considered at all.

In the 1990s, the reducing impact logging (RIL) technique [7, 8] was introduced in the tropical region. It was defined as an "Intensively planned and carefully controlled implementation of harvesting operations to minimise the impact on forest stands and soils, usually in individual tree selection cutting" [9, 10]. The main idea was to adopt a more rational and structured method for forest exploitation. So far, new transaction and overhead cost were introduced in the forest management. More recently, according to the social and environmental relevance of forests, in many countries, forest management is strongly influenced by decisions of forest institution. The main institutions concern the: (a) introduction of forestry and environmental laws and rules; (b) introduction of administrative procedures for safeguarding forest public interest; (c) definition of roles and responsibilities for the economic subjects involved in the transformation process; and (d) redaction of the accidental risks and promotion of high health standards to forest workers.

The main assumptions adopted in the contribution concern economic subjects, operators, and capitals (workers, machines, material, etc.) involved in the transformation processes. Each of them must be paid for the performances provided. Technical and legal aspects are presented as background, while the core is based on the introduction of transformation steps, from being trees of the forest ecosystem to becoming market products. Timber production is one of the ecosystem services of forest ecosystems. Using the analytical approach, costs and revenue have been identified for each technical centrum of expenditure. Dedicated presentation and information have been provided on calculation method and data sources; however, in **Table 1**, a glossary of the main controversial terms has been developed.

2. Legal and technical aspects

2.1 Legal aspects

2.1.1 Public interest on forest ecosystems

Forests provide a large spectrum of ecosystem services that produce well-being to the humanity. Consequently, many institutions currently put silvicultural management under laws and regulations discipline. The target of institutions and administrative proceedings is to safeguard the public interest on forest ecosystems [11].

The following are the two main consequences:

- Introduction of administrative procedures, at the end of which the FLOs obtain the right to perform the silvicultural intervention
- Introduction of transaction costs to prepare the technical documentation necessary to satisfy the administrative process

2.1.2 Entrustment of the silvicultural intervention

Utilization intervention can take place under the regime of the following:

- Sale: with which the property transfers to the LCs the property rights of the forest stand destined to be felled
- Contract: with which the FLOs entrust through the acquisition of LC services, the forest harvesting operations, while they maintain the timber ownership
- Economy: where the FLOs, in particular the public ones, decide to carry out the work in-house, using the own property, in particular machinery, other tools, and forest workers enrolled

Regardless of the regime, the entrustment can be done in terms of the following:

- Purchase all timber at fixed price: LC undertakes to carry out the intervention at the agreed and invariable price respecting the identified unit and following technical indications from the project and the subsequent notes of the competent institutions.
- Based on a fixed price for unit: regime that requires the commitment by the property to pay the intervention at the fixed price per unit (usually volume or area), as well as the commitment by the company to quantify ex-post the volume or the area.
- Mixed, partly defined at fixed price, and partly on the fixed price for unit.
- These aspects are usually clearly stated in the contract. Further mutual fundamental obligations are:
- for the FLOs: to certify that they have fulfilled all procedures necessary to obtain the authorization to carry out the silvicultural intervention and all provisions have been transposed in area (boundaries of the forest area under management have been marked, trees that shall be fell are also marked, etc.). Documents produced and received must be available to the LCs, which will be used at the end for monitoring the correctness of developed activity; and
- for the LCs: to be aware of the territorial unit where they must operate, the characteristics of the stand and the area, as well as the nature of the silvicultural intervention to be implemented. The LCs certify to be fully aware of the technical and technological complexity level of the silvicultural intervention, to be aware of the appropriate methods to carry it out, as well as to have availability of equipment, workers, and materials necessary for its execution in due time.

2.2 Technical aspects

Evaluation processes require relevant technical information such as: (a) the forest areas in which the stand involved in the management activity is located; (b) the timber volume that should be felled; and (c) the type of marketable products that could be obtained.

2.2.1 Entity of the volume removal

If the stand of volume, silvicultural intervention, and felled timber are expressed in cubic meter, among them, the following relation subsists:

$$\mathrm{SoV} = \mathrm{SI} + \mathrm{TH}$$

And solved for SI, it becomes

$$\mathrm{SI} = \mathrm{SoV} - \mathrm{TH}$$

where [*SoV*] is the volume of stand invested in the ground until cutting intervention starts; [*SI*] is the volume of trees left in the ground at the end of the harvesting process; and [*TH*] is the volume of the trees felled and transformed in market goods from the LCs. Silvicultural intervention is indirectly obtained by felling wood volume in excess, given the adopted forest management system. In other words, it is the result of the forest left in the ground after the trees cut by LCs are already in the market. The function is always verified at the time that felling activity starts.

Silviculture proposes different management methods, in relation to the autoecology of the species and forest community, as well as the land characteristics, forest state and type of the previous management, objectives pursued, and infrastructures. The characteristics of each intervention are defined in terms of the following:

- Volume of the stems that must be released
- Characteristics of the trees to be released with respect to the horizontal (territorial distribution) and vertical (stand stratigraphy) plane
- Characteristics of the trees to be released for environmental needs, biodiversity, and other nonproductive functions

Given the total volume that insists in the area just before felling activity starts, the magnitude of the intervention can be quantified in terms of volume to be withdrawn $\pi = \left[\frac{TH}{STM}\right]$ or to be released $\rho = \left[\frac{SI}{STM}\right]$.

Magnitude of treatment	Mechanization level	Workers' qualification	Productiveness	Monetary results		
				Costs	**Revenue**	**Budget performance**
Slight	Low	Low skill	Low	High	Low	Negative
	Intermediate	Medium skill	Moderate	Moderate—high	Low	Negative—variable
Moderate	Intermediate	Medium skill	Moderate—high	Moderate	Medium—good	Positive
	Advanced	Skill	High	Moderate—high	Variable results	Variable
Effective	Advanced	Skill	Very high	Moderate—low	Substantial	Positive
	Intense	High skill	High	Moderate	Medium—good	Variable
Intensive	Intense	High skill	Very high	Low	Substantial	Positive

Source: our elaboration [12, 15].

Table 2.
Orienting economic results for treatment magnitude types.

Intense	**High-tech advanced machines, developed for forest works, capable of performing combined operations (e g., felling-processing, or extraction-processing or bunching-extraction-transport) for the final assortment production.**	
Harvester + Forwarder		
Labor career	10 years	10 years
Productivity	25–80 m^3/h felling and processing	20–90 m^3/h extraction-transport
Market price (VAT excluding)	360,000–600,000 € per harvester	130,000–380,000 € per forwarder
Operative cost	Harvester 2.2–8.5 €/t	Forwarder 15.0–42.0 €/t
Sources	[13–15]	[13–16]
Cable yarder + processor + chipper on forwarder		
Labor career	10 years	10 years
Productivity	6.2–9.95 m^3/h bunching, extraction, and processing by cable yarding and processor	8.0–55.5 m^3/h bunching, extraction, and chipping by forwarder with chipper
Market price (VAT excluding)	150,000–295,000.00 € Cable yarder and processor	180,000–420,000 € per forwarder with chipper

Intense	**High-tech advanced machines, developed for forest works, capable of performing combined operations (e g., felling-processing, or extraction-processing or bunching-extraction-transport) for the final assortment production.**	
Operative cost	Cable yarder and processor 19.00–57.00 €/t	Forwarder with chipper 4.0–16.0 €/t
Sources	[13–16]	[13–17]
Chainsaw		
Labor career	2.5 years	
Productivity	0.5–4.0 m^3/h felling and processing	
Market price (VAT excluding)	750–2000 € per chainsaw	
Operative cost	Chainsaw 6.5–28.0 €/t	
Sources	[13–16, 18, 20]	

Table 3.
Level of mechanization applied in the forest yard: intense. Chainsaw is the machine common in all the levels, but reported only in this table, in order to limit the repetition.

Description	Technologically advanced machines, developed for forest works, capable of performing single operations (e g., felling or extraction or processing)	
Skidder + forest crane or winch		
Labor career	10 years	
Productivity	Skidder equipped with dual arch grapple or winch 3.5–8.5 t/h)	
Market price (VAT excluding)	150,000–330,000.00 € skidder with forestry equipment	
Operative cost	Skidder with forestry equipment 5.6–18.50 €/t	
Sources	[13–16]	
Cable yarder + chipper		
Labor career	10 years	10 years
Productivity	Cable yarder 0.8–5.5 t/h	15.0–65.5 m^3/h movement, load, and chipping
Market price (VAT excluding)	15,000–180,000 € cable yarder	80,000–260,000 € per chipper
Operative cost	Cable yarder 3.5–19.50 €/t	Chipper 3.0–15.5 €/t
Sources	[13–16]	[13–17]

Table 4.
Level of mechanization applied in the forest yard: advanced.

Basing on the size of the withdrawal, it is possible to have a qualitative indication of the intervention and consequently an estimation of expected revenue (**Table 2**). Minor interventions, refereed to particularly small volume withdrawals, don't need a high mechanization level, which determines the growing employment of workers and a modest productivity. The unit costs of the intervention are high, if compared to constant revenues, with a negative balance. Interventions with higher withdrawals allow the use of higher levels of mechanization, lower use of labor, and consequently greater productivity. These conditions lead to a growing budget balance.

2.2.2 Types of forestry yard

There are different types of forestry yard. The first classification is based on 4 mechanization levels (**Tables 3, 4, 5,** and **6**). The former is characterized by high

	Low-tech advanced machines, developed for rural works, with forestry equipment, capable of performing single operations (e g., felling or extraction or processing)	
Rural tractor with forest winch + forest trailer		
Labor career	12 years	12 years
Productivity	Rural tractor with forest winch 1.5–5.0 t/h	Rural tractor with forest trailer 2.2–7.0 t/h
Market price (VAT excluding)	40,000–100,000 € (rural tractor with forest winch)	50,000–120,000 € rural tractor with forest trailer
Operative cost	Rural tractor with forest winch 10.5–26.5 €/t	Rural tractor with forest trailer 9.5–25.0 €/t
Sources	[13–16, 18]	[13–17]
Rural tractor with bins + rural tractor with chipper		

	Low-tech advanced machines, developed for rural works, with forestry equipment, capable of performing single operations (e g., felling or extraction or processing)	
Labor career	12 years	10 years
Productivity	Rural tractor with bins 1.5–5.5 t/h	Rural tractor with chipper 1.0–15.0 t/h
Market price (VAT excluding)	35,000–75,000 € rural tractor with bins	55,000–145,000 € rural tractor with chipper
Operative cost	Rural tractor with bins 7.0–16.0 €/t	Rural tractor with chipper 12.0–36.0 €/t
Sources	[13–16, 18]	[13–15, 17, 19]

Table 5.
Level of mechanization applied in the forest yard: intermediate.

Description	Low technologies and trained animals	
Chainsaw + mules		
Labor career	2.5 years	18 years
Productivity	0.5–4.0 m^3/h felling and processing	Team of 5–7 mules 1.0–5.5 t/h
Market price (VAT excluding)	750–2000 € per chainsaw	8000–16,000 € team of 5–7 mules
Operative cost	Chainsaw 6.5–28.0 €/t	Team of 5–7 mules 15.0–35.0 €/t
Sources	[13–16, 18, 20]	[13–15, 18, 19]
Horses (TPRs)		
Labor career		
Productivity	Skidding extraction with TPR horse 0.7–3.5 t/h	
Market price (VAT excluding)	1500–6000 € per TPR horse	
Operative cost	Skidding extraction with TPR horse 16.5–22.5 €/t	
Sources	[14, 20]	

Table 6.
Level of mechanization applied in the forest yard: low.

investments and high productivity, with decreasing average costs by increasing processed volumes (intense – **Table 3**); the lower mechanization level is characterized by increasing operating costs by decreasing productivity of processes and work (low or based on animal power—**Table 6**). Other mechanization levels are advanced (**Table 4**) and intense (**Table 5**).

The second classification can be based on the type of productions in the forest (or productions at the felling site). There are four logging system classes (**Tables 7, 8, 9**, and **10**), such as the following:

- Full tree logging system (Whole tree harvesting system) (**Table 7**)

- Tree length logging system (**Table 8**)

First timber products	Final timber production
Images	Timber

	Chips

	Firewood

Table 7.
Logging systems: Extraction of full trees (TLS) (branch and top).

- Intermediate (mix) logging system (**Table 9**)
- Short wood logging system (**Table 10**)

Each type of forestry yard is characterized by different cost dynamics:

- Cost for activity in forest: decreasing costs by reducing the work for each tree
- Cost for bunching-extraction: increasing costs by increasing number of logs to be bunched to achieve that volume to make extraction efficient and economically convenient

First timber products	Final assortment
Images	Timber for construction use

Logs at landing site	Timber for minor use

Table 8.
Logging system: Tree length (TLS). Extraction of full stem.

Image	Final timber products
Full stand cross cutting to multiple market assortments 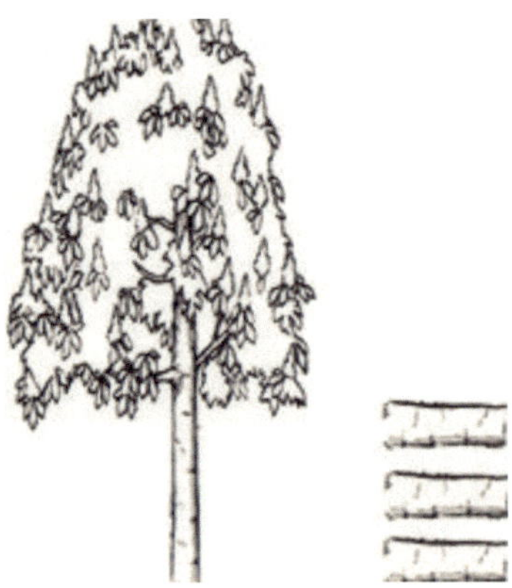	Timber 

Table 9.
Logging system: Intermediate system: Tree length/short wood (cut to length CTL). Extraction full stand cross cutting to multiple market assortments.

Image	
	Fuelwood 
	Chips 

Table 10.
Short wood system (SWS): Wood processed in final assortment in forest.

3. Costs and revenue in the transformation process

3.1 Transformation process

The term "transformation process" refers to the whole process to get stand transformed in row timber material and allocated in the landing, in order to be sold. This process includes all actions that should be done by

- the forest owner, directly by it or indirectly through performance of forest consultant, as in majority of cases and
- the logging company, who develops the technological cycle.

The starting point is the decision assumed by the forest owner to perform silvicultural intervention, while the end is when final monitoring of LCs' activity is done and certification of the results is presented. This process is articulated in four steps, which are as follows:

- Preliminary: aimed to acquire the permit for the silvicultural intervention to be executed.
- Preparatory: which includes (a) the operations to transpose in the forest, the planning, and the administrative provisions; (b) the assignment of the work to an LC; and (c) the signing of the contract between FLO and LC.

Type of costs			Description		Calculation	Details
Land	Structure	Fixed costs	Fee for recovery and storage of machines	QRR	$= [CC * cc_{\%} * A]$	CC = building costs; cc% = recovery and storage coefficient; A = area
Working capital	Machines		Amortization rate	AR	$= \left[\frac{(MP-RV)}{n}\right]$	MP = market price; RV = residual value; n = life time in years
			Interest	Int	$= [QA * r]$	QA = amortization rate
			Maintenance fee	MF	$= [QA * mc_{\%}]$	QA = amortization rate; $mc\%$ = maintenance cost in percent
			Insurance and tax	I&T	$= [\sum (Ins.ce + Tax)]$	Ins.ce = insurance; tax
			Costs of rapid consumption parts	CRCP	$= [\frac{K}{n} * MP]$	K = cost of part; n = life time; MP = market price
	Other tools	Fixed cost	Fee for recovery and storage of machines	QRR	$= [CC * cc_{\%} * Sup]$	CC = building costs; $cc\%$ = recovery and storage coefficient; A = area
			Amortization rate	AR	$= \left[\frac{(MP-RV)}{n}\right]$	MP = market price; RV = residual value; n = life time in years
			Interest	Int	$= [QA * r]$	QA = amortization rate
			Maintenance fee	QM	$= [QA * cm_{\%}]$	QA = amortization rate; $mc\%$ = maintenance cost in percent
			Rapid consumption parts costs	CPRC	$= [\frac{K}{n} * MP]$	K = cost of part; n = life time; MP = market price

DOI: http://dx.doi.org/10.5772/intechopen.87572

Type of costs			Description		Calculation	Details
Material and energy	Fuel and oil	Variable costs	Fuel costs	FK_{Carb}	$= [Kwh * \mu_{um} * C_{sc} * FK\_MP * h]$ or $= [C_{gl} * \mu_{um} * FK\_MP]$	Kwh = power; μ_{um} = technical coefficient; c_{sc} = specific consumption; FK_MP = market price fuel costs; h = hours of use
			Oil costs	OK_{Lub}	$= [Kwh * \mu_{um} * O_{sc} * OK\_MP * h]$ or $= [L_{gl} * \mu_{um} * OK\_MP * h]$	Kwh = power; μ_{um} = technical coefficient; O_{sc} = specific consumption oil; OK_MP = market price fuel costs; h = hours of use
	Consumables		Other material costs	OMC	$= [Q_i * MP_i]$	Q = quantity; MP = market price
Labor	Workers		Salary	S	$= \left[\sum_{i \to 1}^{n} S_i * h_i\right]$	S = salary for hour; h = hours; i = qualification
Financial capital	Financial		Interest on capital advanced	ICA	$= \left[FC_i * r * \frac{t_{(j-s)}}{365}\right]$	FC = financial capital; r = discount rate; t = time in days
Business organization			Entrepreneur remuneration	ER	$= \left[\sum_{i \to 1}^{n} (KT_j) * 10\%\right]$	KT = total costs; j = different cost types

The footnote symbol "" means multiplication.*

Table 11.
Transformation cost types and calculation procedures.

- Executive: in which LC performs silvicultural intervention. It includes selected trees felling, extraction, transportation, and stacking of the timber in the landing.
- Conclusive: in which the goodness of the silvicultural intervention is verified.

3.2 Costs

In order to implement the four steps of the transformation process, many services and inputs must be purchased on the market. Total costs are the sum of three cost types (**Table 10**):

- Transaction costs
- Operative costs
- Overhead costs

3.2.1 Transaction costs

This type of cost includes all expenditures that have to be incurred in the process to fulfill the administrative procedures defined by rules, regulations, and laws, in order to manage the forest. Those expenditures are mainly necessary to acquire the permission to perform the silvicultural intervention, but they include the costs to prepare the forest area, to entrust LCs with the work, to ensure effectively results, and to monitor the results. The whole transaction costs are defined formally frame in which FCs and LCs must work in order to safeguard public interest as well as the forest and other social aspects (work safety aspect). Both FLOs and LCs sustain transaction costs.

FLOs' transaction costs include the following:

- Forest consultancy
- Technical documents and drafts
- Administrative fees
- Selecting and entrusting an LC with implementation of forest utilization
- Technical responsibility to safeguard the FLO's interest during the activity
- Verification of the forest utilization conformity to the standard defined by the permission and the forest rules, regulations, and laws

These costs per unit area usually decrease, but total cost increases according to the area interested in the process (proportional principle) (**Figure 3**). Expenditures for those performances are defined as bunched cost by

- dedicated market survey;
- negotiation between the LC and forest consultant;

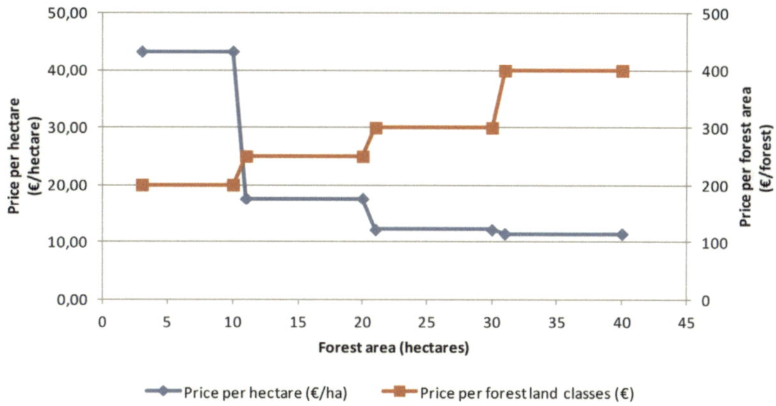

Figure 3.
Transaction costs. Source: Our elaboration.

- tables issued by competent institutions. The amounts are changed proportionally with the volume of timber felled or with the forest area under management;
- mixed approach, combining the two systems mentioned earlier; and
- surveys on the dedicated market.

The LCs' transaction costs are related to the obligations concerning the following:

- Healthy costs, capital investment to ensure high safety standard in the forestry yards and to the forest workers
- Cost of refund deposit, caution money that forest owners can operate when LCs don't pay the stumpage price and for penalty that LCs incurred during the activities

The LC costs are defined by market survey.

3.2.2 Operative costs

All expenditures to implement technological cycle are defined as operative cost (**Table 11**). These costs are sustained to transform trees into marketable timber products, mainly by the LC and only a few for FL. They include expenditure to buy primary and secondary productive factors.

Main primary factors are as follows:

- Land: capital permanently invested in the ground. While it is very relevant for FLOs, for LCs, it is limited to the structures for recovery and storage of working capital, as machineries, machines, and other tools.
- Capitals that include the following:
- Working capital such as machines (forwarder, harvester, chainsaw, etc.) and other tools (winch, etc.). Those have a multiple-year employment, so the use cost must be distributed through the years they will contribute to the activity (amortization costs and annual interest). These costs are added to the

maintenance costs, insurance and other contribution, cost of rapid consumption parts, etc. in order to define the cost machine.

- Financial capital necessary for the possession and use of working capital above descripted, to remunerate workers as well as to cover the interest of financial advances for the activity development.

- Labor, concerning employers involved in the transformation process, such as forest workers and other units that have administrative functions. In both cases, the remuneration changes according to the skill and qualifications. The fundamental information is not the payment for hour or day, but the costs sustained by the entrepreneur that include tax, insurance, and other costs paid as benefits or facilities for the future of workers.

- Entrepreneur organization, done by the person who assumes the forest activity risk. Currently, the payment is split between equity profit and extra profit. The first covers the responsibility assumed by the forest entrepreneur to manage the activity and it is estimated as about 10% of the total cost and named as equity net profit, while extra profit is obtained as differential between total costs (include equity profit) and total revenues. This amount covers the risk management activity.

3.2.3 Overhead costs

Also named as indirect costs, these do not contribute directly to obtain the product but exist to ensure LC functioning. They are related to the LC unit as a whole, and they cannot be applied or traced to any specific unit of output. Overhead costs include the following:

- Costs for managing goods and material purchases involved in the forest utilization process; costs due to the activity of timber trade in the market

- Costs for managing of insurances, taxes, and other contributions due to the LC

- Costs for managing markets and operator networks

- Costs for the LC accounts

3.3 Revenues

The output of the silvicultural activity can be expressed in terms of volume or value. The first results by measurement operations of the standing, with special emphasis to the volume of row timber material removal from the ground [*V*] differentiated for market destination [*i*] and expected to be sold in the polder (first competitive timber market). The latter is the result of the volume of row timber material for the relative market price [*MP*]. Data can be collected by market survey, or dedicated statistical publication.

Formally,

$$VdM = \sum_{j \to 1}^{m} (V_i * MP_i)$$

DOI: http://dx.doi.org/10.5772/intechopen.87572

4. Monetary evaluation of forest management

Main questions of forest management are:

- What is the standing forest market price?
- What is the timber forest products' market price?
- What is the forest management cost?

Literature offers three approaches to answer the questions above, which are as follows:

- Market price approach
- Cost approach
- Combination of the above-mentioned approaches

In all these approaches, results are based on common comparative method. This method ensures a strictly direct or indirect connection between market and the good under evaluation.

4.1 Appraisal theory

Theoretical background on evaluation method has been defined from International Valuation Standard Council [21]. In this contest, two main approaches are suitable: market comparison approach and cost approach. The first obtains the timber value by comparing the timber under evaluation with other similar timbers sold in the market; for which, price and at least one technical parameter are well known. The latter defines the value considering all expenditures that the enterprise have to sustain in order to obtain the product under evaluation.

Timber evaluation can have different assessments depending on whether one of the following two objectives is pursued:

a. Timber optimization uses: FLOs' target is to ensure the most appreciated market product.

b. Optimization of market functioning: FLOs target to create the most favorable conditions for large market participation by the LCs.

The first target tends to favor the major LCs, even if the LCs who would take part in the market are very few in number, at least only one. Those LCs are technologically advanced, have greater financial availability, and have wider timber markets, as they can be international timber markets. The hypothesis is that this setting should ensure an effective use of timber and that it can achieve the highest addend value. On the other hand, the second target tends to align itself with the most frequent conditions compared to the local area framework, so as to allow the greatest participation of the local LCs at the market. The hypothesis is that if a large number of LCs take part in the market, that should ensure highest LC competition and the highest stumpage price.

4.1.1 Market value

4.1.1.1 The stumpage value

The first step is to acquire an adequate observation numbers, at least not less than 4 for each variable used in model, of

- market price (dependent variable);
- technical variables (independent variables),

the value of the stand can be determined through two procedures:

a) By direct comparison, using the fundamental proportion to evaluate a market good. Having market prices and at least one technical parameter value, the proportion adapted to evaluate forest stand marketable is

$$\sum_{i \to 1}^{n} SV_i : \sum_{i \to 1}^{n} Vol_i = SV_x : Vol_x$$

where [*SV*] is the stumpage price, [*Vol*] is the volume felled, [*i*] is the number of market observations collected by a survey, and [x] are the data related to the stand under evaluation. Developing the proportion above in favor of SV, it becomes

$$SV_x = \left[\frac{\sum_{i \to 1}^{n} SV_i}{\sum_{i \to 1}^{n} Vol_i} \right] * Vol_x$$

where the ratio in the square bracket is the stumpage value for cubic meter.

b) By indirect comparison, building an econometric model

$$Y = f(x_1, \dots, x_i, \dots, x_n, \varepsilon)$$

where [*Y*] is the dependent variable vector of the stumpage price, [*x*] are the generic technical variables, and [*i*] is the type of variables such as forest area (hectares), timber volume (cubic meter), infrastructure index (qualitative data), and other parameters.

The strong limits of both procedures are (a) the lower number of LCs that take the risk that low number of LCs have an informal agreement about the stumpage value, and the LC that acquire the stem it was decided before the timber market start officially; and (c) there isn't a well structured culture on how and what forest data, technical and market, should be collected. Each forest owner has its collection, and each forest owner itself selects the variables that should be registered.

4.1.1.2 Timber raw material market

It is the market in which the trees, transformed in marketable products, are sold as timber raw material. That market has two relevant advantages for the evaluation proceeding: (a) even if the number of FLOs or LCs that support the supply is very low, the sawmills are much more so the market should have less distortion; and (b) it is the first market later to the stand felling.

The market price of timber raw material is obtained by market survey. Database is built using the price registered in the market.

4.1.2 Forest management evaluation costs

The production cost approach concerns the technological cycle step and it includes only the expenses necessary to carry it out. This circumstance mainly happens when forest management has social objectives or the timber raw material has high market value and the FLOs prefer it to be sold directly in the timber raw material market because they expect strong completion among sawmills. The FLOs operate on service markets. They purchase the LC services for felling, processing and transportation of plant to an area which is easy to access (landing). Timber raw material can be

- evaluated to reduce forest management costs;
- made available to the local community to pursue their objectives;
- sold in the timber raw material, directly or through dedicated agencies.

FLOs have to pay the LCs in any case. Activity is developed within the regulation code of "tender" to fell stand. The characteristic of this entrust is that FLOs have to pay LCs for their performance. The main national law states that an enterprise, as LCs, "assumes, (...), the fulfilment of a work or a service towards a consideration in money"[1].

Production cost is formally obtained as

$$K_{Tot} = \left[\sum_{i \to 1}^{n} k_i\right] + \left(k_i * r * \frac{t_{(j-s)}}{365}\right)$$

where $[K]$ is the total costs, $[k]$ is the elementary costs, $[i]$ is the types of costs, $[r]$ is the discount rate, and $[t]$ time and $[j, s]$ are, respectively, the day when the work finished and the day when the expenditure has been done.

4.1.3 The transformation value

The last procedure provides the evaluation of the stand as a comparison between the value of the timber raw material market and all costs necessary to transform the stand into marketable products. The transformation process increases timber value step by step until it becomes timber raw material. The evaluation process, on the other hand, moves in the opposite direction: starting from the market products to achieve the stumpage value (**Figure 4**).

The fundamental relationship at the base of the procedure is that timber raw material market price is equal to the sum of stumpage price with the costs of carrying out transforming process:

$$MP_{TRM} = SV + K_{Tot}$$

where $[MP_{RTM}]$ is timber raw material market price in the first market after the stand is felled; [i] is the types of timber product obtained (timber construction, fuel wood, etc.); $[SV]$ is the stumpage price; $[K_{Tot}]$ is the total costs of transformation

[1] Civil Code, article 1655.

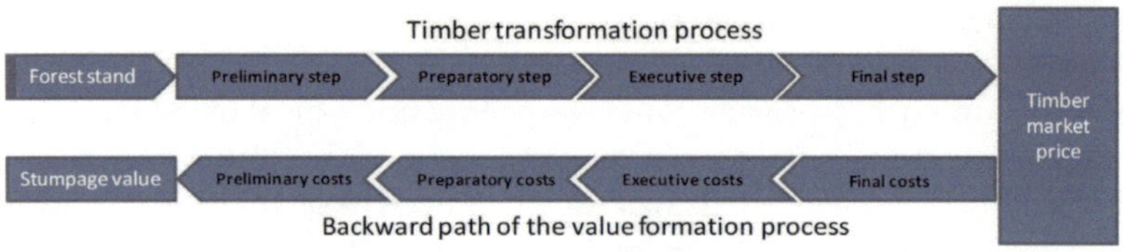

Figure 4.
Pathway of timber productions and value formation.

processes from stand to timber raw materials; and [j] is the types of costs. Resolving for the stumpage value, it becomes

$$SV = [MP_{TRM} - K_{Tot}] = \left[\left(\sum_{i \to 1}^{n} MP_{TRMi} \right) - \left(\sum_{j \to 1}^{m} K_{Totj} \right) \right]$$

The ex ante budget is the tool that foresters usually adopt, where in one site is reported the revenue and in the other site the expenditures. The balance between revenues and expenditures is the stumpage value that LCs take from the commitment to pay at the FLOs when agreement was signed.

4.2 Appraisal approach for entrusting types

4.2.1 Forest management in house

The common model of this forest management is based on the ability of forest property (public or private) to carry out the forestry intervention. The owner directly or through an agency of the same subsidiary carries out forest utilization using personnel, machines, and tools in its possession. The economic and financial questions that accompany this approach are as follows:

- What is the total cost of carrying out the intervention (K_{tot})?
- What is its operating cost (K_{Op})?
- What is the market value of the timber raw material (MV_{TRM})?

Figure 5.
Services market: forest land owner purchase logging company service for felling, extraction, and processing timber production at lowest price. Source: Our elaboration.

The total cost expresses the total amount of costs regardless of the evidence that the resources used are internal. The operating cost focuses only on the variable (additional) costs that are incurred only if the intervention is carried out, ignoring the costs related to the internal resources involved in the works and the costs that the property still support. The last question relates to the value of timber raw material, which is quantified through market surveys.

4.2.2 Forest management by tender

In this case, the silvicultural intervention is entrusted to an LC, which provides a service to the FLOs in exchange for payment of the service. The company that carries out the intervention is the one that, all other parameters being equal, ensures the service at the lowest price (**Figure 5**).

	A	B	C	D	E	E formula	G
1	Types of data	Description	Range	Units	Amounts		Sources
2	Economic data	Market price		€	45,000.00		Market survey
3		Percentuale di recupero		%	10.00%		Technical documents
4		Value at the end of the career		€	4500.00	=E2 × E3	Our elaboration
5		Annual amortization		€	4050.00	=(E2 − E4)/E10	Our elaboration
6		Market price tires		€	2000.00		Market survey
7		Gasoline price		€/l	1.12		Market survey
8		Discount rate		%	3.00%		Market survey
9		Average annual investment		€	27,000.00	=E2 × E18	
10	Technical data	Economic duration		year	10.00		Technical documents
11		Annual machine usage hours		hours	1000.00		Technical documents
12		Technical duration		hours	10,000.00		Technical documents
13		Work days in the year		days	240.00		Technical documents
14		Working days in hours		hours	4.20		Technical documents
15		Power		HP	80.00		Technical documents
16		Tires duration		hours	3000.00		Technical documents

	A	B	C	D	E	E formula	G
17		Gasoline consumption duration per hours		l/ hour	6.15	=(E20 × E15 × E22)/ E19	Our elaboration
18	Coefficients and parameters	Average annual investment coefficient		%	60.00%		Technical documents
19		Tires coefficient			1.20		Technical documents
20		Gasoline conversion coefficient (1 l = 0.84 kg)	0.70–0.85		0.84		Technical documents
21		Gasoline consumption			0.17		Technical documents
22		Lubricants consumption		%	10.00%		Technical documents
23		Load factor	0.38–0.70		0.38		Technical documents
24		Maintenance coefficient	100–30%	%	100.00%		Technical documents
25		Variable expenditures coefficient	15–5%	%	7.00%		Technical documents
26	Fixed costs	Annual amortization per hour		€/ hours	4.05	=E5/E11	Our elaboration
27		Interests per hours		€/ hours	0.81	=E9 × E8/E11	Our elaboration
28		Variable expenses		€/ hours	1.89	=E9 × E24/E11	Our elaboration
29		**Total fixed costs**		**€/ hours**	**6.75**	**=E25 + E26 + E27**	**Our elaboration**
30	Variable costs	Maintenance and repair costs per hour		€/ hours	4.05	=E5/E11 × E23	Our elaboration
31		Gasoline cost per hour		€/ hours	6.89	=E17 × E7	Our elaboration
32		Lubricant cost per hour		€/ hours	0.69	=E30 × E21	Our elaboration
33		Tires cost per hour		€/ hours	0.80	=E19 × E6/E16	Our elaboration
34		**Total variable costs**		**€/ hours**	**12.43**	**=E29 + E30 + E31 + E32**	**Our elaboration**
35	**Total costs**	**Total machine costs**		**€/ hours**	**19.18**	**=E28 + E33**	**Our elaboration**

Source: our elaboration on frame [4, 5].

Table 12.
Cost machines using the FAO frame.

	A	B	C	D	E	E explosed	G
1	Types of data	Description	Range	Units	Amounts	Formulas	Sources
2	Economic data	Market price		€	45,000.00		Market survey
3		Percentuale di recupero		%	10.00%		Technical documents
4		Value at the end of the career		€	4500.00	=E2 × E3	Our elaboration
5		Annual amortization		€	4050.00	=(E2 – E4)/E11	Our elaboration
6		Labor cost		€/ hours	14.00		Market survey
7		Tire market price		€	2000.00		Market survey
8		Gasoline price		€/l	1.25		Market survey
9		Lubricant price		€/l	2.25		Market survey
10		Discount rate		%	3.00%		Market survey
11	Technical data	Economic duration		year	10.00		Technical documents
12		Annual machine usage hours		hours	1000.00		Technical documents
13		Technical duration		hours	10,000.00		Technical documents
14		Work days in the year		days	250.00		Technical documents
15		Working days		hours	8.00		Technical documents
16		Hours machine usage effectively		hours	4.20		Technical documents
17		Power		kW	60.00		Technical documents
18		Tire duration		hours	3000.00		Technical documents
19	Coefficients and parameters	Gasoline conversion coefficient (1 L = 0.84 Kg)	0.70–0.85		0.84		Technical documents
20		Oil conversion coefficient (1 L = 0.95 kg)	950–850		0.95		Technical documents
21		Gasoline specific consumption	280–300	g/ kWh	300.00		Technical documents
22		Oil specific consumption	2–4	g/ kWh	4.00		Technical documents

	A	B	C	D	E	E explosed	G
23		Use coefficient of available power	95–50%	%	66.67%		Technical documents
24		Maintenance and repair coefficient	0.10–0.13	%	0.13		Technical documents
25	Variable expenditures coefficient	2.5– 0.5%	%	2.50%		Technical documents	
26	Fixed costs	Annual amortization per hour		€/hours	4.05	=E5/E12	Our elaboration
27		Interests per hour		€/hours	0.61	=(((E2 − E4)/2) × E10)/E12	Our elaboration
28		Variable expenses		€/hours	1.13	=(E2 × E25)/E12	Our elaboration
29		**Total fixed costs**		**€/hours**	**5.78**	**=E26 + E27 + E28**	**Our elaboration**
30	Variable costs	Maintenance and repair costs per hour		€/hours	1.82	=E6 × E24	Our elaboration
31		Gasoline cost per hour		€/hours	9.38	=1/E19 × (E21/1000) × E23 × (E16/E15) × E17 × E8	Our elaboration
32		Oil cost per hour		€/hours	0.30	=1/E20 × (E22/1000) × E17 × (E16/E15)*E9	Our elaboration
33		Tires cost per hour		€/hours	0.67	=E7/E18	Our elaboration
34		**Total variable costs per hours**		**€/hours**	**12.16**	**=E30 + E31 + E32 + E33**	**Our elaboration**
35	**Total costs**	**Total cost machines per hours**		**€/hours**	**17.94**	**=E29 + E34**	**Our elaboration**

Source: our elaboration on [6].

Table 13.
Cost machine per hour, using the USDA frame.

Different approaches to calculate operating costs are reported in the cost machine literature. The main frames are elaborated and reported in **Tables 12** and **13**, respectively, for FAO and USDA. The following are the common comments:

- Transaction and overhead costs are not included.
- Labor costs are not included.
- The frame proposed is developed for machine that works in huge areas or regions.
- Total costs have to be used as an approximation of the cost machines.

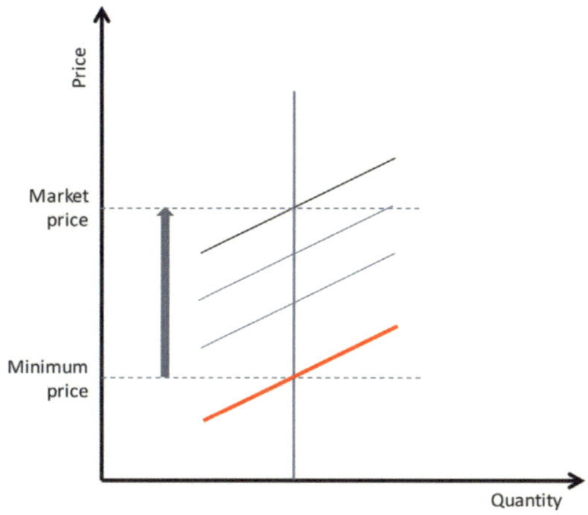

Figure 6.
Commodities market: forest land owner sells its stand to the logging company that makes the highest price.

- Some algorithms and parameters used for evaluating cost are not easy to understanding the economic ratio.

4.2.3 Forest management by sale of stand

In the forest appraisals, the approach is to elaborate an ex ante budget of the silvicultural intervention including the expected costs to transform trees in marketable products and the expected revenue that should be obtained from the products sold.

The differences between expected revenue and costs are the expected stumpage value of the trees that LCs should pay to the FLO to bay the stand, while only the amount of the expected costs is the price that FLO has to pay to the LC for the service of felling the stand. Stumpage price became the minimum price that FLO accepts to sell its stand. LCs that want to purchase it have to submit a proposal with a price higher than the minimum (**Figure 6**).

5. Conclusions

Growing awareness of the usefulness of forest ecosystems makes the operational cost significant as a component of the wider transformation cost. The latter includes both transaction costs in order to satisfy the provisions dictated by the legislative and regulatory forest and related forest disciplines, as well as the overhead costs that allow the correct functioning of the LCs.

Approaches introduced by international institutions lend themselves to an assessment, very approximate of the costs of managing uniform forests that cover large and flat areas. Their limits are given by concentrating on the component of operating costs, excluding overhead and transaction costs, as well as the introduction of simplifications in order to increase the territorial scale of application. They determine an underestimation of forest management costs [22].

A drawback instead overcomes the analytical approach, whose strong point is its adaptation to the context of intervention and to the specificities of the transformation cycle. This makes it possible to overcome the deformities that characterize forests, especially in the mountain areas, where it is possible to register a different stumpage value for two similar forests, close to each other and having the same productions. This approach also ensures transparency and traceability of the assessment process, as well as flexibility being able to be adapted to the different process for entrusting the management of the stand.

Acknowledgements

This research was in part supported by the "Departments of Excellence—2018" Program of the Italian Ministry of Education, University and Research (Law 232/2016), financed. Department for Innovation in Biological, Agro-Food and Forest Systems (DIBAF)-University of Tuscia, Project "Landscape 4.0—Food, Well-being and Environment" and Department of Agriculture and Forest Science (DAFNE), University of Tuscia, Project (WP3).

Author details

Francesco Carbone[1]* and Rodolfo Picchio[2]

1 Department for Innovation in Biological, Agro-Food and Forest systems (DIBAF), University of Tuscia, Italy

2 Department of Agricultural and Forestry Science (DAFNE), University of Tuscia, Italy

*Address all correspondence to: fcarbone@unitus.it

References

[1] Bettinger P, Siry JP, Boston K, Grebner DL. Forest Management and Planning. Academic Press: Elsevier; 2017

[2] Grebner DL, Bettinger P, Siry JP. Introduction to Forestry and Natural Resources. Academic Press: Elsevier; 2012

[3] UN. 2015

[4] FAO. Cost Control in Forest Harvesting and Road Constraction. Roma: FAO; 1992

[5] FAO. Cost Control in Forest Harvesting and Road Construction. Food and Agriculture Organization of the United Nations; 1992, Forestry paper, n. 99

[6] Miyata ES. Determining fixed and operating costs of logging equipment. General Technical Report NC-55. North Central Forest Experiment Station, USDA Forest Service; 1980

[7] Marsh CW, Tay J, Pinard MA, Putz FE, Sullivan TE. Reduced impact logging: A pilot project in Sabah, Malaysia. In: Schulte A, Schöne D, editors. Dipterocarp Forest Ecosystems: Towards Sustainable Management. Singapore: World Scientific Publishing Co. Pte. Ltd; 1996. pp. 293-307

[8] Putz FE. Approaches to sustainable forest management. In: Working Paper No. 4. Bogor: CIFOR; 1994. pp. 7

[9] Putz FE, Sist P, Fredericksen T, Dykstra D. Reduced-impact logging: Challenges and opportunities. Forest Ecology and Management. 2008;**256**: 1427-1433

[10] Sist P. Reduced impact logging in the tropics: Objectives, principles and impacts. International Forestry Review. 2000;**2**(1):3-10

[11] Carbone F. Institution, forest enterprises and transaction costs on the domestic market. Journal of Agricultural Economics. 2012;**1**:89-121

[12] Hippoliti G. Appunti di Meccanizzazione Forestale. Società Editrice Fiorentina: Firenze, Italy; 1997

[13] Verani S, Sperandio G, Picchio R. First thinning in a coniferous plantation for biomass production: productivity and costs. In: Proceeding FORMEC 2010 Forest Engineering: Meeting the Needs of the Society and the Environment; 11-14 July 2010; Padova, Italy. 2010. ISBN 978 88 6129 569 8. Available from: http://www.tesaf.unipd.it/formec2010/Proceedings/Ab/ab100.pdf-

[14] Verani S, Sperandio G, Picchio R, Savelli S. La raccolta della biomassa forestale. Tecniche, economia e sicurezza sul lavoro. Vol. 1. Monterotondo (Roma): Grafica Salaria; 2009. p. 50

[15] Verani S, Sperandio G, Picchio R, Spinelli R, Picchi G. Field Handbook—Poplar Harvesting; Poplar Harvesting. International Poplar Commission Working Paper IPC/8. Forest Management Division. Rome: FAO; 2008. pp. 54

[16] Marchi E, Neri F, Fabiano F, Cambi M, Picchio R. Pianificazione, organizzazione e gestione delle utilizzazioni forestali per la prevenzione selvicolturale. In: Bovio G, Corona P, Leone V, editors. Gestione selvicolturale dei combustibili forestali per la prevenzione degli incendi boschivi. Arezzo: Compagnia delle Foreste; 2014

[17] Picchio R, Sirna A, Sperandio G, Spina R, Verani S. Mechanized harvesting of eucalypt coppice for biomass production using high mechanization level. Rivista Croatian

Journal of Forest Engineering. 2012; **33**(1):15-24

[18] Picchio R, Spina R, Maesano M, Carbone F, Lo Monaco A, Marchi E. Stumpage value in the short wood system for the conversion into high forest of a oak coppice. Rivista Forestry Studies in China. 2011;**13**(4):252-262. DOI: 10.1007/s11632-013-0411-7

[19] Civitarese V, Sperandio G, Picchio R. Aspetti economici della produzione di cippato; capitolo di manuale tecnico: Processi di valorizzazione del cippato agroforestale PRO.VA.CI. AGR. Viterbo; 2015. p. 108, ISBN: 979-12-200-0444-2-

[20] Picchio R, Antogiovanni A, Calienno L, Caputo F, Marziali L, Venanzi R, Lo Monaco A. Utilizzazioni e meccanizzazione forestale; capitolo di libro: Progetto MORINABIO, L'Aquila, 2015. 134 p. ISBN: 978–88-95453-26-2

[21] International Valuation Standard Council. International valuation standard 2013. Framework and requirements. London; 2013

[22] Piegai F, Fratini R, Pettenella D. Costi macchina, confronto fra diversi metodi di calcolo, 2008. Sherwood—Foreste ed Alberi Oggi. Aulla Magna; no. 8. 2008. p. 27

Chapter 6

Raw Material Demand-Supply and Policy Recommendations of Turkish Wood-Based Panel Industry

Hasan Tezcan Yildirim

Abstract

The wood-based panel industry is one of the fast developing and growing sectors in the world. As of the year 2017, Turkey is the fourth biggest wood-based panel producer with a share of 3.9%. The fast sectoral development is considered as a positive indicator, although unplanned growth is not desirable. In this scope, the raw material Turkey possesses, and the opportunity to meet the future demand of the sector has been investigated. The estimated production capacity of Turkey for the year 2018 is calculated as 6,657,294 m^3/year for particle board using two average alternative models. The sector's possible yearly demand concerning the production capacity is approximately 11–12 million m^3 besides the 8–9 million m^3 production from the local production import gain ground. Providing a solution concerning the raw material supply, increasing the industrial afforestation, amplifying the state aid in the local products, and taking the necessary measures in order to decrease the cost is crucial. The said measures might have a significant role to offer a solution for the problems of the sector. The future projections should aim at reaching a solution to the raw material problem and the technical problems.

Keywords: fiberwood, fuelwood, demand, supply, forestry, Turkey

1. Introduction

The development in the forest industry has progressed to the use of solid-like materials rather than solid wood materials. The main reason behind this progress is thought to be the demand and the concern on the capability of meeting this demand. Although wood is an important raw material, it has become more difficult to meet the demand in every passing year [1]. Because the formation of wood-based raw material obtained from forests needs a quite long period of time. The growth of wood-based industries all over the world has made the use of new substitute materials instead of wood inevitable [2–4]. On the other hand, neither the diversity in substitute materials nor the use of both wood-based materials and other materials at the same time has reduced the demand for wood. In this respect, the wood industry is subject to a constant development and change [5]. The wood-based panel industry is an important forest-based one in China. At this point, for example, wood-based panels have high economic importance in China economy. Some projections show that the production of the wood-based panel industry has expanded considerably in recent years and is expected to increase with an average annual growth rate of 1.05% from 2015 to 2030 [6, 7]. One of the most leading sectors in terms of this mentioned change and development has been the board industry.

In fact, the forest industry has been an essential leading branch of the industry since old days in terms of social development [8–12]. Therefore, the use of wood in the industry continues as it was in the past. Similarly, the use of wood as an industrial material is still an important source of income for both those who produce wood and treat wood to produce wood products [13, 14]. When the subject is approached in terms of the industry, another reality is that the particleboard and fiberboard industries have developed particularly in recent years and a heavy raw material demand exists [15]. When the issue is considered from a historical perspective, the industrial production of particleboard started in 1941 in Germany and showed a rapid development after 1948 [16]. Although the fiberboard industry started in the early 1900s, the large-scale commercial production emerged between the two world wars in the United States [17–19]. The main reason for the fact that the particleboard industry first emerged and developed in Central European countries is the desire to substitute wood with a new and more economical construction material with more convenient dimensions and to make savings from wood use just like the other construction materials as a result of the destruction caused by the Second World War [20].

The fact that the private sector completely dominated the forest product industry since the early 1990s has been accompanied by huge investment in the field by private companies. When considered from this point of view, it can be stated that the wood-based panel industry has carried out a great development in the past 20 years in particular [21]. However, rapid growth and development have brought adverse effects as well. The difficulty in meeting raw material demand comes at the top of these adverse effects. The most serious bottleneck in meeting the demand for raw material is considered to be the prices and the amount of the demand [12, 22, 23]. Public dominated production in Turkey has usually been a problem in meeting the raw material demand of the private sector. This situation has led the sector to import, but as a result of the recent changes in the raw material exporting policies of the countries and the economic events, this option has become insufficient in solving the problem.

The estimated production capacity of the board industry in Turkey is 5.1 million m^3 of particleboard per year. As for fiberboard, the production amount is 6.8 million m^3 per year [24]. The total number of production facilities in the sector is 35 of which 19 produce particleboard and 16 produce fiberboard. The total production capacity of the industry is approximately 12 million m^3 per year, whereas the actual production is 8.6 million m^3 per year (Particle Board Industry Association [25]). Therefore, in addition to particleboard and fiberboard purchased by the sector from General Directorate of Forestry (OGM), fuelwood has also been added to the demand list of the sector. Latest investments in the fiberboard and particleboard industries and capacity enhancement attempts are expected to move the sector further. However, the expectation of low raw material supply for the sector is assumed as the biggest obstacle for the companies by the representatives of the sector against production enhancement (particularly in 2013). The expectations by the industry, the production amount of the forestry organization, and the changes in related policies shall directly affect the future of the industry. At this point, it is crucially important to estimate the changes in the raw material supply of the wood-based industry in the forthcoming period.

This study aims to introduce suggestions on meeting the raw material demand, which is considered to be the main problem of the wood-based panel industry. The raw material supply amount of OGM, which is the main raw material supplier for the industry, has been projected by considering the particle-fiber wood, which is an important kind of raw material for the sector, and fuelwood production amount between the years 1977 and 2017. Certainly, the presence of various social and economic factors has been taken into consideration while making the projection. The main factors taken into consideration might be listed as follows: population; gross national product per capita; and afforestation fields, which are important for the sustainability of the forestlands and the unit sale price of the particle-fiber wood and fuelwood. One of the important variables to be considered in terms of the results of the study is the number of companies operating in the board industry and their production capacities. However, these variables have not been able to be evaluated under findings as there are not any regular statistics on the issue, but they have been evaluated in the suggestions provided under the conclusion part instead. The fact that no data can be found on the particle-fiber wood production of OGM until the year 1977 has been effective in gathering data starting from the year 1977.

2. Material and method

As it is known, the dependent variable in an economic event is sometimes affected by a single independent variable and sometimes by more than one independent variable. When the dependent variable is explained by more than one independent variable, multidimensional decision-making methods are used [26]. Multidimensional decision-making methods are suitable for the structure of forest resources, and with the use of the method, more significant decisions and solution offers in forest resources management can be created [27, 28]. In this respect, multidimensional decision-making methods are of the most frequently used methods in forestry studies. Regression analysis is one of the appropriate multidimensional decision-making methods for the study.

In regression modeling, the intended use has to be well defined in order to find the most appropriate regression model [29–31]. Since the long-term data of the previous years (between 1977 and 2017) had been obtained regularly on a yearly basis and the purpose was to estimate the raw material production to meet the demand, regression modeling has been preferred to use. Two techniques are used in regression analyses. They are simple regression analysis and multiple regression analysis. Multiple linear regression modeling has been determined as the most appropriate modeling technique for the study as it provides the chance of evaluating multiple data. Multiple linear regression analysis has been formed for the purposes of revealing how the production amount of particle-fiber wood and fuelwood changes depending on the specified independent variables and determining the raw material amount that can be provided by country resources for the industry. Future projections concerning supply and demand equilibrium have been made regarding the established capacity of the industry (taking into account the available quantitative data range and the data quantity) as well. In terms of research technique, Durbin Watson (DW) statistic and coefficient have been utilized first, in order to test the autocorrelation among the independent variables used in the multiple linear regression analysis.

While determining the particle-fiber wood and fuelwood supply amounts of the wood-based panel industry according to the data by OGM, particle-fiber wood production amount (Y1) and fuelwood production amount (Y2) have been specified as dependent variables. Unit sale price of the particle-fiber wood (X1), unit sale price of fuelwood (X2), afforestation rate (X3), population (X4), and current producer prices in the US dollar basis with the gross national product (X5) have been accepted as independent variables. The data related to the mentioned variables have been derived from the databases of OGM, Turkish Statistical Institute (TÜİK), İstanbul Chamber of Commerce (ITO), State Planning Organization (DPT), and the World Bank. The data including the number of facilities in the industry and

Years		Particle-fiber wood productions (m^3)	Fuel wood productions (m^3)	Particle-fiber wood unit prices (\$USD/m^3)	Fuel wood unit prices (\$USD/m^3)	Afforestation (ha)	Population	Gross national product per person (GNPP) (\$USD)
		Y_1	Y_2	X_1	X_2	X_3	X_4	X_5
1977	1	171,000	20,309,000	15.24	7.70	37,985	41,316,300	1427
1978	2	184,000	20,071,000	14.84	8.72	34,050	42,206,200	1550
1979	3	173,000	20,046,000	19.66	10.85	27,867	43,132,600	2079
1980	4	164,000	21,949,000	18.63	11.25	20,969	44,347,719	1564
1981	5	180,000	20,192,000	12.75	10.99	45,943	45,130,000	1579
1982	6	439,000	20,372,000	12.06	8.35	53,680	45,353,405	1402
1983	7	742,000	19,851,000	16.06	7.48	66,210	46,965,156	1310
1984	8	953,000	16,659,000	22.38	8.46	87,627	48,735,507	1247
1985	9	884,000	14,289,000	20.83	12.52	100,400	50,664,458	1368
1986	10	1,071,000	12,138,000	15.08	11.05	108,354	51,706,684	1511
1987	11	913,000	12,503,000	23.59	8.18	114,132	52,770,350	1706
1988	12	1,137,000	12,942,000	21.16	12.72	119,369	53,855,897	1745
1989	13	1,193,000	13,062,000	15.20	11.14	113,639	54,963,775	2022
1990	14	1,113,000	12,145,000	17.20	13.47	78,884	56,473,035	2794
1991	15	1,104,000	11,503,000	15.69	12.70	56,752	57,512,139	2736
1992	16	1,177,000	11,146,000	29.50	17.19	24,519	58,570,362	2842
1993	17	1,004,000	10,846,000	36.25	25.44	27,058	59,648,057	3180
1994	18	1,363,000	8,379,000	16.60	10.94	39,652	60,745,581	2270
1995	19	1,320,000	9,539,000	19.46	13.59	24,257	61,863,300	2898
1996	20	1,362,000	10,402,000	32.75	20.20	37,927	63,001,585	3054
1997	21	1,406,000	9,246,000	20.04	14.88	32,031	64,160,814	3144
1998	22	1,278,000	8,372,000	24.17	15.97	25,959	65,341,373	4497
1999	23	1,252,000	8,167,000	21.73	14.25	11,529	66,543,654	4108
2000	24	1,371,209	7,861,442	21.64	14.43	24,494	67,803,927	4317
2001	25	1,254,599	7,576,683	15.10	9.38	25,672	68,064,972	3120
2002	26	1,821,253	7,506,725	22.91	13.61	28,647	68,327,022	3660
2003	27	2,073,150	7,815,932	34.83	22.10	36,914	68,590,081	4718
2004	28	2,329,897	8,119,555	38.67	24.96	34,016	68,854,153	6041
2005	29	2,409,446	7,667,026	42.51	26.85	21,439	69,119,242	7384
2006	30	2,964,647	7,003,026	41.23	25.85	25,319	69,729,967	8035
2007	31	3,265,092	6,834,024	47.64	27.66	18,228	70,586,256	9710
2008	32	3,816,522	7,303,889	52.59	31.71	39,467	71,517,100	10,851
2009	33	4,033,257	7,427,596	41.37	25.86	46,872	72,561,312	9036
2010	34	4,608,171	7,194,372	43.32	27.99	41,857	73,722,988	10,672
2011	35	4,662,578	6,778,101	44.31	26.95	39,964	74,724,269	11,341
2012	36	5,424,794	6,432,674	51.88	36.82	42,009	75,627,384	11,720
2013	37	5,551,397	5,981,703	43.13	27.35	46,656	76,667,864	12,543
2014	38	6,608,416	5,257,995	38.39	25.60	40,325	77,695,904	12,127
2015	39	6,866,355	5,022,986	34.56	21.32	38,986	78,741,053	10,985
2016	40	7,201,462	4,877,067	34.09	19.86	48,230	79,814,871	10,863
2017	41	6,494,372	4,359,646	29.06	20.29	46,935	80,810,525	10,541

Table 1.
Depended and independent variables used in the study and their values.

their production capacity have also been obtained from the databases of the aforementioned institutions and from their reports related to the sector. All the obtained data are given in **Table 1**.

While carrying out the analyses, first of all, the changes in the independent variables (X1, X2, X3, X4, and X5) according to years have been tested using mathematical methods, a method of time-series analysis. In other words, mathematical formulas that represent the correlation and then the probability model have been reached by using the diagram that shows the correlation among the dependent and independent variables. The correlation among all the variables has been tested by means of diagrams in the study. Thus, estimated independent variable values that are to be used in explaining the values of the dependent variables in the upcoming years have been obtained at the first stage. At the second stage, multiple linear regression analysis was use. Two alternative models have been exploited for the purpose of numbering the estimations. In the first alternative model, all the independent variables were integrated regardless of reliability. Also, in the second alternative model, the reliability, which is below 0.05, was integrated, and thus, the model was constituted. The result of two alternative models was presented, and the resulting difference in number was put forward. Finally, the resulting differences were evaluated to be whether neglected or not. The numeric data analyses have been carried out using the software SPSS.

3. Findings and discussion

The data between the years 1977 and 2017 have been analyzed through this method with the help of the software SPSS. The results, the formulas for each regression model, and R^2 values are given in **Table 2**. The coefficient for the variable "year" has been assumed as 1 for the initial year, which is 1977, and 54 for the year 2030. Therefore, a 54-year trend has been composed with the study.

As it can be seen in **Table 2**, except for the variable "afforestation rate," an increase trend is estimated for all variables. After estimating the possible future values of the independent variables, models related to the production amount of particle-fiber wood and fuelwood, which are dependent variables, have been formed using multiple linear regression modeling. Multiple linear regression modeling has been preferred because all variables show a linear relation. **Table 3** indicates not only the results of multiple linear regression analysis, which has been used to estimate the production amount of particle-fiber wood and fuelwood, but also the models obtained and the independent variable coefficients involved in the models.

When the values of the dependent variable "particle-fiber wood and fuelwood production amount" (Y_{1A}) in **Table 3** are observed, it can be understood that 94.4% of the dependent variable (Y_{1A1}) in Model 1 is explained by the independent variables involving in the model. The remaining 5.6% is explained by the variables that are not involved in the model due to the term "error." As for the second model, it is understood that 91.5% of the dependent variable (Y_{1A2}) is explained by the independent variables involving in the model, and the remaining rate is explained by the variables that are not involved in the model. In this case, it can be concluded that the variables picked for the model are highly effective. It is understood from the DW test scores that autocorrelation does not exist in estimating the particle-fiber wood production in the first and second models of which DW test scores are 0.961 and 0.441, respectively in **Table 3**. On the other hand, both the first model, where the modeling is significant at every level as a whole (F = 118.659/Sig = 0.000) and the second model (F = 133.369/Sig = 0.000) can be stated to be significant (Significance = Sig).

Years	Estimation of particle-fiber wood unit sale price ($USD/m^3)	Estimation of fuel wood unit sale price ($USD/m^3)	Estimation of afforestation (ha)	Estimation of population	Estimation of gross national product per person (GNP) ($USD)
R^2	0.58	0.64	0.45	0.99	0.73
Formula	X_{1A} = 0.6395×Year+15.24	X_{2A} = 0.4794×Year +7,7	X_{3A} = −0.3067×Year4 + 33.711×Year3–1157.9×Year2 + 12,306×Year +37,985	X_{4A} = 986.603 × Year +4E+07	X_{5A} = 205.28 × Year +1427
2018 (42)	42.10	27.83	55,525	81,437,326	10,049
2019 (43)	42.74	28.31	57,900	82,423,929	10,254
2020 (44)	43.38	28.79	59,851	83,410,532	10,459
2021 (45)	44.02	29.27	61,261	84,397,135	10,665
2022 (46)	44.66	29.75	62,003	85,383,738	10,870
2023 (47)	45.30	30.23	61,945	86,370,341	11,075
2024 (48)	45.94	30.71	60,947	87,356,944	11,280
2025 (49)	46.58	31.19	58,862	88,343,547	11,486
2026 (50)	47.22	31.67	55,535	89,330,150	11,691
2027 (51)	47.85	32.15	50,804	90,316,753	11,896
2028 (52)	48.49	32.63	44,499	91,303,356	12,102
2029 (53)	49.13	33.11	36,444	92,289,959	12,307
2030 (54)	49.77	33.59	26,454	93,276,562	12,512

Table 2.
Estimated values of independent variables through regression modeling.

Dependent variables	(R^2)	*F*	Sig.	Durbin Watson (DW) test	Independent variables and parameters			
					Independent variables	Values	*t* Test	Sig.
First alternative model results								
Particle-fiber wood production (Y_{1A1})	0.94	118.65	0.00	0.961	Constant	−1,970,429.679	−2.449	0.019
					X_{1A}	−26,787.535	−0.996	0.326
					X_{2A}	−50,531.489	−1.179	0.246
					X_{3A}	10.652	3.387	0.002*
					X_{4A}	0.039	2.651	0.012*
					X_{5A}	575.574	10.385	0.000*
Fuel wood production (Y_{2A1})	0.96	172.49	0.00	1.062	Constant	48,623,723.470	27.881	0.000
					X_{1A}	−35,917.345	−0.642	0.525
					X_{2A}	−29036.430	−0.326	0.746
					X_{3A}	−17.267	−2.641	0.012*
					X_{4A}	−0.551	−18.719	0.000*
					X_{5A}	631.647	5.396	0.000*
Second alternative model results								
Particle-fiber wood production (Y_{1A2})	0.92	133.36	0.00	0.441	Constant	−2,832,784.647	−3.74	0.014
					X_{3A}	12.556	3.402	0.002
					X_{4A}	0.038	2.162	0.037
					X_{5A}	427.350	8.331	0.000
Fuel wood production (Y_{2A2})	0.96	274.20	0.00	0.758	Constant	49,774,524.600	27.881	0.000
					X_{3A}	−15.708	−2.641	0.012
					X_{4A}	−0.574	−18.719	0.000
					X_{5A}	492.363	5.396	0.000

**Statistically significant at ≤0.05.*

Table 3.
Projection modeling results related to particle-fiber wood and fuelwood production.

Two alternative models that have been formed according to the coefficients obtained from multiple linear regression analysis in order to estimate particle-fiber wood production are given in Eqs. (1) and (2) as follows:

$$Y_{1A1} = -1,970,429.679 - (26,787.535 \times X_{1A}) - (50,531.489 \times X_{2A}) + (10.652 \times X_{3A}) + (0.039 \times X_{4A}) + (575.574 \times X_{5A}) \quad (1)$$

$$Y_{1A2} = -2,832,784.647 + (12.556 \times X_{3A}) + (0.038 \times X_{4A}) + (427.350 \times X_{5A}) \quad (2)$$

With the help of the model formed, particle-fiber wood production average of Turkey has been estimated to be 6,657,294 m^3 for 2018 (average 1 and 2 models). According to the OGM [32] records, the production was 7,131,469 m^3 at the end of October 2018. When the estimated and actual production amounts are compared, a difference of 474,000 m^3 can be seen, which means an error margin of 6% meaning that the reliability of the projection has been proved with the rate of 94%. **Table 4**, on the other hand, indicates the estimated amounts covering the years between 2018 and 2030 by both of the alternative models and quantitative difference between the models.

Years	Y_{1A} (first model)	Y_{1A} (second model)	Difference between estimates (Y_{1A1}–Y_{1A2})	Estimates percentage error (%)
2018	7,056,371	6,258,217	798,154	11.31
2019	7,238,005	6,433,786	804,218	11.11
2020	7,415,121	6,604,032	811,090	10.94
2021	7,586,465	6,767,472	818,993	10.80
2022	7,750,702	6,922,535	828,167	10.69
2023	7,906,417	7,067,554	838,863	10.61
2024	8,052,121	7,200,771	851,350	10.57
2025	8,186,243	7,320,337	865,906	10.58
2026	8,307,135	7,424,307	882,828	10.63
2027	8,413,070	7,510,648	902,422	10.73
2028	8,502,244	7,577,231	925,013	10.88
2029	8,572,773	7,621,836	950,937	11.09
2030	8,622,694	7,642,150	980,544	11.37

Table 4.
The estimated amount of fiber-particle wood production 2018–2030.

On the other hand, when the dependent variable "fuelwood production amount" (Y_{2A}) is examined in **Table 3**, we come to the result that 96.1% of the dependent variable (Y_{2A1}) is explained by the independent variables involved in the first model. The remaining 8.5% is explained by the variables that are not involved in the model due to the term "error." As for the second model, it is understood that 95.7% of the dependent variable (Y_{2A2}) is explained by the independent variables involving in the model, and the remaining rate is explained by the variables that are not involved in the model. In this case, it can be concluded that the variables picked for the model are highly effective. As seen in the DW test scores, autocorrelation does not exist in estimating the fuelwood production in the first and second models of which DW test scores are 1.062 and 0.758, respectively in **Table 3**. On the other hand, both the first model, where the modeling is significant at every level as a whole (F = 172.491/Sig = 0.000) and the second model (F = 274.206 / Sig = 0.000) can be stated to be significant (Significance = Sig).

The models related to the projection of fuelwood production according to the coefficients obtained through multiple linear regression analysis are given in Eqs. (3) and (4) as follows:

$$\begin{aligned} Y_{2A1} &= 48,623,723.470 - (35,917.345 \times X_{1A}) - (29,036.430 \times X_{2A}) \\ &\quad - (17.267 \times X_{3A}) - (0.551 \times X_{4A}) + (631.647 \times X_{5A}) \end{aligned} \tag{3}$$

$$Y_{2A2} = 49,774,524.600 - (15.708 \times X_{3A}) - (0.574 \times X_{4A}) + (492.363 \times X_{5A}) \tag{4}$$

With the help of the model formed, fuelwood production average of Turkey has been estimated to be 5,957,586 m^3 for 2018 (average 1 and 2 models). According to the OGM [32] records, the production was 5,866,939 m^3 at the end of October 2018. When the estimated and actual production amounts are compared, a difference of 90,647 m^3 can be seen, which means an error margin of 1.5% meaning that the reliability of the projection has been proved with the rate of 98.5%. **Table 5**, on the other hand, indicates the estimated amounts covering the years between 2018 and 2030 by both of the alternative models and quantitative difference between the models.

Years	Y_{2A1} (first model)	Y_{2A2} (second model)	Difference between estimates (Y_{2A1}–Y_{2A2})	Estimates percentage error (%)
2018	5,815,096	6,100,077	−284.982	−4.01
2019	5,302,710	5,577,000	−274.290	−4.15
2020	4,797,647	5,060,584	−262.937	−4.31
2021	4,301,941	4,552,681	−250.739	−4.46
2022	3,817,757	4,055,259	−237.501	−4.62
2023	3,347,385	3,570,402	−223.016	−4.77
2024	2,893,243	3,100,309	−207.066	−4.90
2025	2,457,874	2,647,295	−189.420	−4.99
2026	2,043,952	2,213,791	−169.839	−5.02
2027	1,654,274	1,802,342	−148.068	−4.95
2028	1,291,768	1,415,611	−123.844	−4.72
2029	959,485	1,056,376	−96.891	−4.24
2030	660,606	727,528	−66.922	−3.38

Table 5.
The estimated amount of firewood production 2018–2030.

According to the calculations by DPT [33] related to the energy and fuel required for production, 1.9 m^3 wood is required for the production of 1 m^3 particleboard. As stated in Section 1, in the event that the established capacity of the industry remains constant, approximately 10.2 million m^3 of wood per year shall be required for the production of 5.1 million m^3/year of particleboard. Similarly, according to the calculations by DPT in 2007, 1.2 m^3 wood is required for the production of fiberboard. Under the circumstance that fiberboard production capacity remains constant, approximately 8.2 million m^3 of wood per year shall be required for the production of 6.8 million m^3 per year, and the total need for wood shall be 18.4 million m^3 for a full-capacity production. When the capacity of the factories is kept at 80%, the amount shall be 14.7 million m^3. According to the projections made in the study, the possibility of meeting this amount under these conditions is not considered to be favorable. Because the average annual production is 11–12 million m^3 in the short term, whereas the long-term production decreases to 8–9 million m^3 following the decrease in fuelwood production.

4. Conclusion

In the light of findings obtained within the scope of the study, it has been clearly found out that the raw material supply for wood-based panel industry may turn into a problematic issue. Considering that the supply of raw material shall mainly be provided by the OGM, the raw material problem shall begin to increase within the next 20 years. Several studies carried out on the issue present similar conclusions ([4, 23, 34, 35]). At this point, the sectors where panel products are used are also of great importance. Construction sector and furniture industry are the leading ones among those sectors. Since the study focuses on raw material supply, the demands on a sectoral basis have not been discussed. The following suggestions, on the other hand, have been suggested as a result of the findings by approaching the board industry as a whole:

Raw material supply is one of the most important issues affecting the structural development of the sector. Therefore, demand projection should be made, and these projections should be revised on a yearly basis in order to provide the sector with a sustainable growth through accurate planning. Diversification of supply sources for balancing sectoral demand shall be the most important policy change as well. Considering the economic balances, not only the planning by the sector but also the involvement of the state, which holds 99.9% of the country's forest assets, in this planning shall be crucial. Besides, the state should increase the incentive opportunities.

Although the projection for raw material production does not point out any serious problems for the present but alarms for the possible ones in the future. Because the actual average production of particleboard and fiberboard is 11–12 million m^3 per year. The raw material demand for such a production is calculated to be 18 million m^3. While the current demand by the sector is hardly met, a greater bottleneck shall be created with a reduction in the fuelwood production amounts. From the point of view, in addition to the need to keep fuelwood production amount constant, it may be appropriate for OGM to focus on the production of wood that meeting the needs of the sector.

Another problem that Turkey might encounter in providing the sector with raw material is the fact that wood-based energy generation emerges with the energy agenda of the country. Particularly, the countries' tendency toward wood-based energy generation as an alternative way in order to provide the security of supply is likely to create a new kind of raw material bottleneck. The board industry, which has a slight chance of competing with the energy sector in raw material supply, is expected to encounter problems such as a shrink or capacity slow down. Moreover, the fact that energy forestry does not become widespread in the country seems to cast a shadow over the sector in the short term rather than the long term.

Based on the projections that sectoral demand shall increase and new conditions of competition shall occur, it is of great importance not only for the private sector but also for the state to engage in afforestation activities using fast-growing species, particularly around the factories with great production capacities.

As stated above, different alternatives or new policies may be identified for the solution of the raw material problem. One of these possible solutions might be the prioritization of the practices that are important to particularly meet the quantitative wood demands of the forest industry with a silvicultural technique. Here it is possible to consider the expansion of the afforestation using fast-growing species such as red pine or a reduction in the management period.

Another important policy might be the designation of areas for the production of wood within the framework of functional planning, implementation of these plans, and reviewing the forest management plans. At this point, a policy to be followed might be the expanding the forestlands designated for wood production.

Author details

Hasan Tezcan Yildirim
Faculty of Forestry, Department of Forest Engineering, İstanbul University
Cerrahpaşa, İstanbul, Turkey

*Address all correspondence to: htezcan@istanbul.edu.tr

References

[1] FAO, 2018: Global Demand for Wood Products. http://www.fao.org/docrep/pdf/011/i0350e/i0350e02a.pdf (Visited date: 20 November 2018)

[2] Lata K, Dubey B, Misra AK. Modeling the effects of wood and non-wood based industries on forestry resources. Natural Resource Modeling. 2016;**29**(4):559-580

[3] Souza AM, Nascimento MF, Almedia DH, Lopes Silva DA, Almedia TH, Christoforo AL, et al. Wood-based composite made of wood waste and epoxy based ink-waste as adhesive: A cleaner production alternative. Journal of Cleaner Production. 2018;**193**:549-562

[4] Yıldırım HT, Candan Z, Korkut S. Wood-based panels industry in Turkey: Future raw material challenges and suggestions. Maderas Ciencia y Tecnologia. 2014;**16**(2):175-186

[5] Loučanová E, Palus H, Dzian M. A course of innovations in wood processing industry within the forestry-wood chain in Slovakia: A Q methodology study to identify future orientation in the sector. Forests. 2017;**8**(210):1-13

[6] Kun Z, Wenming L, Hashiramoto O. Demand and Supply of Wood Products in China. FAO Forest Products Working Paper 1; 2007

[7] Wang S, Zhang H, Nie Y, Yang H. Contributions of China's wood-based panels to CO_2 emission and removal implied by the energy consumption standards. Forests. 2017;**8**(273):1-16

[8] İlter E, Ok K. Marketing Principles and Management in Forestry and Forest Industry. Ankara, Turkey: Improvement Second Press. Form Ofset Bookstore; 2007

[9] Mahapatra A, Mitchell CP. Sustainable development of non-timber forest products: Implication for forest management in India. Forest Ecology and Management. 1997;**94**:15-29

[10] Ok K. Idea marketing in forestry: Some implications from the Turkish forestry experience. Forest Policy and Economics. 2005;**7**:493-500

[11] Paul S, Chakrabarti S. Socio-economic issues in forest management in India. Forest Policy and Economics. 2011;**13**(1):55-60

[12] Yıldırım HT. Turkish Wood Based Panels Industry: Future Challenges and Suggestions. Proceedings of the 55th International Convention of Society of Wood Science and Technology, August 27-31, 2012 - Beijing, China; 2012

[13] Cubbage FW, Newman DH. Forest policy reformed: A United States perspective. Forest Policy and Economics. 2006;**9**:261-273

[14] Damette O, Delacote P. Unsustainable timber harvesting, deforestation and the role of certification. Ecological Economics. 2011;**70**(6):1211

[15] Hosseini M, Brege S, Nord T. A combined focused industry and company size investigation of the internationalization-performance relationship: The case of small and medium-sized enterprises (SMEs) within the Swedish wood manufacturing industry. Forest Policy and Economics. 2018;**97**:110-121

[16] Jochem D, Janzen N, Weimar H. Estimation of own and cross price elasticities of demand for wood-based products and associated substitutes in the German construction sector. Journal of Cleaner Production. 2016;**137**:1216-1227

[17] Bozkurt AY, Göker Y. Chipboard Industry İ.Ü. Publication Number: 3614,

İ.Ü. Faculty of Forestry Publication Number: 413, İstanbul; 1990

[18] Erol YS. Differences between urban and rural population with respect to demand on forestry aspects, in a case study of the Turkish province of Balıkesir. Ciência Rural, Santa Maria. 2012;**42**(3):436-443

[19] Suchsland O, Woodson GE. Fiberborad Manufacturing Practices in the United States. United States Department of Agriculture Forest Service Agriculture Handbook No: 640; 1987

[20] Huey BM. Problems of timber products procurement during world war II, 1941-1945. Graduate Student Theses. Dissertations, & Professional Papers. 3314; 1951

[21] Garcia R, Freire F. Environmental assessment of wood-based panels: A comparison of life-cycle-based tools. International Journal of Sustainable Construction. 2012;**1**:2182-2743

[22] Gerasimov Y, Seliverstov A. Industrial round-wood losses associated with harvesting systems in Russia. Croatian Journal of Forest Engineering. 2010;**31**(2):111-126

[23] Yıldırım HT. Examination of wood production-consumption relations in terms of Forest policy in Turkey, doctoral thesis. Istanbul, Turkey: Istanbul University Institute of Science; 2010 (in Turkish)

[24] AIMSAD. 2017. The World's 5th Largest Producer of Turkey Attacks Grows Through New Investments in the Sector Plate. https://www.aimsaddergisi.com/dunyanin-5-buyuk-ureticisi-turkiye-levha-sektoru-yeni-yatirimlarla-buyume-ataginda/ (Visited Date: 12.10.2018)

[25] Particle Board Industry Association, 2017. Statistical database 2012, Istanbul (in Turkish)

[26] Oubedda L, Erraha B, Khalfaoui. Multidimensional analysis data to create a decision support system dedicated to the university environment. Global Journal of Computer Science and Technology Software & Data Engineering. 2012;**12**(13):10-16

[27] Daşdemir İ, Güngör E. Multivariate decision-making methods and their using areas in forestry. The Journal of ZKÜ Bartın Faculty of Forestry. 2002;(4):1-19

[28] Guarini MR, Battisti F, Chiovitti A. A methodology for the selection of multi-criteria decision analysis methods in real estate and land management processes. Sustainability. 2018;**10**(507):1-28

[29] Altunışık R, Coşkun R, Yıldırım E, Bayraktaroğlu S. Sosyal Bilimlerde Araştırma Yöntemleri SPSS Uygulamalı, Sakarya Kitabevi. İkinci Baskı, Sakarya; 2002. ISBN 975-8644-07-6

[30] Yazıcıoğlu, Y. ve Erdoğan S. Spss uygulamalı bilimsel araştırma yöntemleri. Ankara: Detay Yayıncılık; 2004

[31] Kalaycı S. SPSS Applied Multivariate Statistical Techniques. Ankara: Second Press, Asil Bookstore; 2006, 975-9091-14-3

[32] OGM. 2018. Official Forestry Statistics https://www.ogm.gov.tr/ekutuphane/Sayfalar/Istatistikler.aspx?RootFolder=%2Fekutuphane%2FIstatistikler%2FOrmanc%C4%B1l%C4%B1k%20%C4%B0statistikleri&FolderCTID=0x012000301D182F8CB9FC49963274E712A2DC00&View={4B3B693B-B532-4C7F-A2D0-732F715C89CC} (Visited time: 20 November 2018)

[33] DPT. 9th Development Plan (2007-2013). Specialized Commission Report on Wood Products and Furniture. Ankara; 2007

[34] Aras U, Kalaycıoğlu H. Wood Based Composites and Application Areas. Uluslararası Hakemli Mühendislik ve Fen Bilimleri Dergisi. Sayı; 2016. p. 6

[35] Başyiğit C, Çankıran O, Taş HH. The raw materials used in the production of chipboard and the use of wooden waste materials for this aim. SDÜ Journal of Science Institute. 2000;**4**(1):26-31

Chapter 7

Timber Harvesting Production, Costs, Innovation, and Capacity in the Southern Cone and the U.S. South

Patricio Mac Donagh, Joshua Roll, George Hahn and Frederick Cubbage

Abstract

We performed research in the Southern Cone of South America and in North Carolina USA that examined logging production, costs, innovation, and capacity. We compare the findings of this timber harvesting research up until 2015 between South America and the U.S. South, and draw conclusions regarding comparative forestry sector economic advantages. Logging production rates per firm have increased, reaching as much as 200,000 tons per year in the U.S. South, and more than 300,000 tons per year in the Southern Cone. Average total costs for logging were generally less in the Southern Cone, at less than $10 per ton for cut and load at roadside for transport, and more than $12.50 per ton for cut and load in the U.S. South. Logging firm innovation usually led to greater production and reduced costs, and focused mostly on improved timber harvesting systems and processes and use of firm performance monitoring, software, and training. Logging sector capacity was a concern in the U.S. South given aging owners and workers, and most likely to come from expansion by existing firms. The Southern Cone had better prospects to expand logging operations due to higher production rates and more favorable rural worker attitudes toward logging employment. Overall, logging production rates will increase; average total costs are apt to remain relatively stable; innovation will focus on system improvements and management skills such as measurement and monitoring; but capacity for sufficient in the woods and transport workforce will be a continuing issue.

Keywords: forest harvesting, logging, productivity, capacity, innovation, Southern Cone, USA South

1. Introduction

Global competition for wood supply is intense, and it is expected to increase more in the future in order to meet the growing demand for wood fiber [1]. Increased demand places continued pressure on the world forests, and more effective utilization of forests is one of the best and most immediate responses to contribute to forest conservation. The timber and forest products industry value chain begins with standing timber in forests and extends to a variety of manufactured forest products. Contract and independent timber harvesting and transport firms comprise the key link in the supply chain to move timber products from the woods to the mill.

Timber harvesting efficiency and utilization improvements are reflected in reduced costs, and comprise a large share of delivered wood and mill costs for forest products firms. For example, a survey by Hahn [2] in North Carolina found that logging costs averaged 35% of the total wood costs delivered to a mill, with trucking adding another 18%, stumpage 39%, and wood dealer/procurement costs 8%. This chapter focuses on analyses of the comparisons of logging productivity and cost component of the overall timber production, harvesting, and transport value chain in the Southern Cone of South America and the U.S. South.

The different labor, harvesting and transportation systems' costs, coupled with the fluctuation of exchange rates, have led to different levels of supply costs in different countries over time [3]. These harvesting operation components complement differential yields from fast growing forest plantations, which favor South America. These integrated timber growth and harvesting supply chains make it possible to more effectively satisfy demand, reduce the rising real cost of wood, and conserve global forests [4].

The timber and forest products industry value chain begins with standing timber in forests and extends to a variety of manufactured forest products. Private logging firms are key links in this supply chain and may range from independent contractors who buy and harvest timber without long term contracts, to companies with long term contracts with forest products manufacturing facilities, to company crews that work for small or medium size sawmils. Timber logging firms have often been characterized as having high capital requirements, with modest profits per unit of wood harvested. This necessitates both efficiency and high volumes of output to offset high equipment fixed costs. In the Southern Cone of South America, logging firms have expanded from small scale local operations to more sophisticated capital-intensive operations as the forest industry has expanded greatly in the last few decades. In the U.S. South, mechanization occurred sooner, and continues apace, but scarce labor and capital availability have created concerns about the viability of the logging force throughout the Americas.

1.1 Objectives

The objective of this chapter is to summarize findings from recent research that we have completed and other relevant literature on timber harvesting (e.g., logging) production and costs in the Southern Cone of South America and in the U.S. South, in order to examine the contribution of logging to overall forestry sector comparative advantage between the two regions. We cover our empirical research on timber harvesting in the Southern Cone and in North Carolina in the U.S. South, and buttress this with extensive literature on logging throughout both broad regions.

The Southern Cone comprises the countries of Brazil, Uruguay, Argentina, and Chile; the U.S. Southern forests consist of 13 states running from Texas in the west to Virginia in the East. These two regions combined produce more than 20% of total global roundwood production as of 2008, with about 180 million m^3 of wood produced in 2008 in each region compared to the global total of 1.5 billion m^3 [5]. The Southern Cone industrial roundwood production increased about 25% since then [6], while the U.S. South remained fairly constant [7].

Global forest industry profits were decreased by the U.S. housing crisis and the general economic recession of 2007, and the pressure on profits and long run logging contractor supply has been problematic. Manufacturing capacity in the U.S. and in the Southern Cone has since rebounded, but concerns remain that a reduced logging force could hinder forest industry expansion in the Americas.

In order to assess future prospects for timber harvesting capacity and innovation in the Americas, we will examine logging productivity, costs, capacity, and innovation in the Southern Cone and the U.S. South using mixed methods of theory and principles, case studies, and literature synthesis. This will include a broad overview of logging firm productivity and costs in the Americas; a detailed analysis of production and innovation in the Southern Cone; and a simulation of the amount of capacity needed to meet projected increases in wood fiber production the state of North Carolina, USA. Conclusions regarding logging productivity, capacity, innovation, and prospects will be drawn from this synthesis.

1.2 The wood supply chain

The U.S. wood supply chain has undergone significant changes over the past 30 years. Intensive forest management increased forest productivity dramatically. Logging businesses, logging business owners, and their equipment have changed as well. Logging businesses rapidly mechanized their operations during the 1960s–1980s, and the logging industry transformed from a labor-intensive to a capital-intensive industry [8].

South America has substantially expanded its forest plantations and raw material supply. From 1997 to 2005, South America had a high annual growth rate in the production of industrial roundwood, with Brazil and Chile being the most important countries [9]. South America also has the fastest growing industrial timber plantations in the world, comprised mostly of exotic softwood species from the United States and eucalyptus from Australia [10].

From 1997 to 2005, Asia had the only negative regional production growth rate in the world, and China became the largest roundwood importer in the world [9]. The forest industry has grown consistently in recent years, and much of this growth have been focused in the Southern Cone countries of Brazil, Chile, Argentina, and Uruguay. Like any global commodity industry, forest production and harvesting are driven by costs. The development and competitiveness of companies have been based on planted forest and timber harvesting factors like low production costs, excellent plantation growth, and the availability of large areas for afforestation [10, 11].

Independent timber harvesting companies are the vital component of the wood supply chain that harvest timber on public and private forestland and deliver it to forest products mills. Without logging businesses, gains in forest productivity cannot be captured and the chief advantage of forestland investments, biological growth, could not be monetized well [8].

1.3 Timber harvesting production, costs, and innovation

Timber harvesting average total costs depend on the productivity and cost of the individual factors of production such as feller-bunchers or harvesters, skidders or forwarders, loaders or chippers, trucks, and labor. In addition, management skill, entrepreneurship, innovation, safety, infrastructure of roads and government, and environmental protection measures affect average total harvesting costs. Various principles and literature address these factors that affect timber harvesting productivity and capacity.

For the most part, independent logging contractors perform most of the logging now in the Southern U.S. and in the Southern Cone. In 1960s–1970s in the Southern U.S., many large integrated forest products firms had their own logging crews, but they were consistently more expensive than similar independent crews, due to less productivity and higher average wages than independent crews, and the need to pay higher costs for some social insurance than small operators. Thus all major U.S. pulp and paper firms phased out of the logging business by the 1980s. Some separate wood dealers still do maintain contract logging crews, as do some small sawmills. This trend to independent loggers for large forest products firms also occurred in the Southern Cone.

The independent logging capacity has increased through the last 25 years, mainly based in more and bigger equipment. For example, 20 years ago, a common skidder had a 5 tons payload, and nowadays has increased to more than 10 tons. In the case of forwarders 20 years ago a 10 tons loading capacity was common, and nowadays goes up to 20 tons. In the same way, capital invested in machinery increased; 20 years ago, a standard equipment mix demanded about US$ 500,000, and today needs more than US$ 1.5 million [8, 11].

As firms have increased their capital investment, average production per dollar invested has declined. Annual production per $1000 invested declined from 200 tons in 1987 to 140 tons in 2012 [8]. Loggers must remain profitable to remain in business and continue investing in their businesses. Therefore, the logging and forest products industries must innovate in order to remain competitive globally.

Technology innovation in forestry has been described as following paradigm shifts and discontinuous evolution. Technology innovations in general are also referred to as technology change, technology shift, and technology development. The *Oslo Manual* [12] provides one of the most comprehensive definitions of innovation because it allows for consideration of new products, processes, markets, and organizational methods with respect to an individual firm. Then because of an innovation a novel device or method is offered to the market (technology push) or market needs trigger innovations (market pull). Drivers for the innovation process can be either internal or external.

Lindroos et al. [13] describe three main drivers of harvesting mechanization: new technology, new products, and new rules. They argue that irrespective of the size of jumps in technological advances, those expected over the next few decades will most likely be seen as fine tuning of current timber harvesting operations. Stone et al. [14] surveyed 13 logging contractors in Maine (USA) to assess innovation. Based on the Oslo manual, they characterize innovation in four types: (1) Product, (2) Process, (3) Organizational, and (4) Marketing. Product innovation consists of introducing a significantly improved good or service. Process innovation focuses on a significantly improved production system. Organizational consists of a new method in the firm's business practices, workplace organization or external relations. Marketing involves significant changes in product design, placement, promotion, or other strategies. In Maine, they found that logging innovators can and will engage in four types of innovation, but process and product innovation dominate [14].

Externally driven process innovations were most common according to the Maine study, such as new equipment or system configurations, machine computer or GPS applications, or new high flotation tires and tracks for equipment. These generally focused on increasing profitability through reduced cost per unit of production. Product innovations, such as road maintenance, power line maintenance, or specialty harvests, were more common than new products, such as biomass or firewood. Organizational innovation was less common, but controlled more by innovations within the firm. This included better information gathering and analysis by the firm, including computer programs, customized tracking systems, and targeted efficiency improvements. Innovations in marketing their services were used the least by firms, and then mostly in areas with multiple small tracts. However, four firms did note that they had logging certification with one of the two major systems, which did help them get access to markets that noncertified loggers could not [14].

Drawing from OECD and Eurostat [12], Stone et al. [14] provide a useful schematic of the logging innovation system in Maine or elsewhere (**Figure 1**).

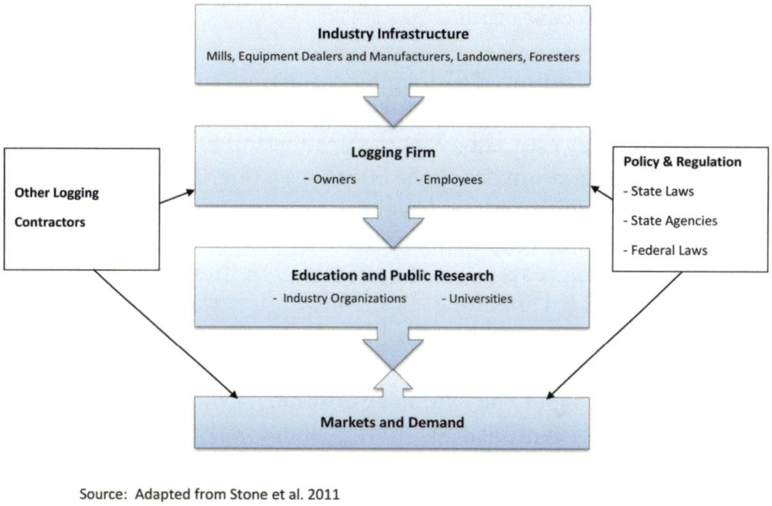

Figure 1.
Logging innovation system in Maine [14].

The external industry infrastructure—mills, equipment manufacturers and dealers, landowners, and foresters—has the most direct influence on logging innovation. Internal drivers are of course the owners and employees of the firm. Firms are also influenced by other logging contractors, and business and environmental policy and regulation, including state and federal laws and state agencies. Education and public research were posited to affect innovation, but not examined in the Maine case studies. Finally, markets and demand provided feedbacks to firms, instigating innovation.

2. The Southern Cone

We analyzed timber harvesting production, costs, and innovation in the Southern Cone of South America based on an extensive survey conducted by the senior author of this chapter, which provided details for the region. In the Southern Cone, forest harvesting activities are carried out through logging contractors that have emerged through the phenomenon known as outsourcing. This is the most common approach for a company that owns forest plantations. Many contractors have been attracted to the logging industry in the Southern Cone, but have failed, because they are not able to maintain both the requirements of the contracting company, and their own company's profitability. It can be inferred that much of the successful contractors could be in practice employees under contract to the larger companies, or in other cases, leading entrepreneurs, who have been able to innovate in the business, and thus develop agile and flexible companies that make a successful long-term business relationship [15].

To meet the increases in demand, to lower logging costs, to reduce environmental damage, or to achieve or maintain levels of global competition, the introduction of technology was one key driver in expansion of the forest products sector in the Southern Cone. For much of the 1980s and the 1990s, the introduction of technology occurred through machinery investment, and from about 1990, technology has included investment in hard technologies (equipment) and soft ones (training, computer programs). Increasingly more companies incorporate soft technologies in timber harvesting, which will increase in the future. However, the largest total investment amount still is for machinery.

2.1 Firm production and cost models

Mac Donagh et al. [11] surveyed 67 logging companies in Argentina, Brazil, and Uruguay, which comprised a substantial 20% of the total timber harvest production in the three countries. They estimated logging production and cost averages and functions for each of those countries in the Southern Cone. **Table 1** summarizes the average production per month and total average costs per ton for the Southern Cone by country, species, contractor type, and mechanization level. Argentina had the lowest average total logging costs (US$ 7.41 per ton), but they were not significantly different than Brazil (US$ 8.41 per ton). Costs in Uruguay were significantly greater at $14.81 per ton. Brazil and Uruguay had the highest average production levels, albeit not significantly greater than Argentina due to the considerable variability in the sample.

Another notable finding of the extensive Southern Cone research was that average total costs did not vary significantly among pine, eucalyptus, or mixed species harvests, nor did production per month. Logging firms that had contracts with pulp mills averaged three times as much production per month, but had about the same costs as those who logged for sawmills. Production levels for fully mechanized firms were more than five times greater per month than for semi-mechanized firms. Average production levels were greatest for mixed harvest types; second for final harvests; and least for thinnings. However, average total costs were not much different, and in fact cheapest for thinnings, but none of these averages were statistically different.

Variable/mean	n	Average total costs (US$/ton)		Average production (ton/month)	
		Cost	Standard error	Production	Standard error
Region					
Argentina	22	7.41	0.89	13,616	9167
Brazil	17	8.19	0.71	26,291	7268
Uruguay	10	**14.82**	1.32	25,660	13,597
Species					
Pine	28	8.35	0.92	16,827	14,317
Eucalyptus	30	9.31	0.89	22,643	8117
Both	9	9.44	1.63	36,211	7842
Contractor type					
Sawmills	29	8.63	0.90	9402	7725
Cellulose	38	9.15	0.79	**31,676**	6748
Harvest type					
Thinning	14	8.43	0.74	11,607	6520
Final harvest	43	9.53	1.30	22,045	11,427
Both	10	10.22	1.53	36,590	13,521
Mechanization level					
Semi-mechanized	20	8.82	0.71	5423	9314
Fully mechanized	47	9.18	1.09	**29,104**	6076

Numbers in bold signify statistically significant differences per Tukey HSD at p = 0.05.

Table 1.
Selected results of an analysis of variance of logging firms in the southern cone, 2011–2012 survey data.

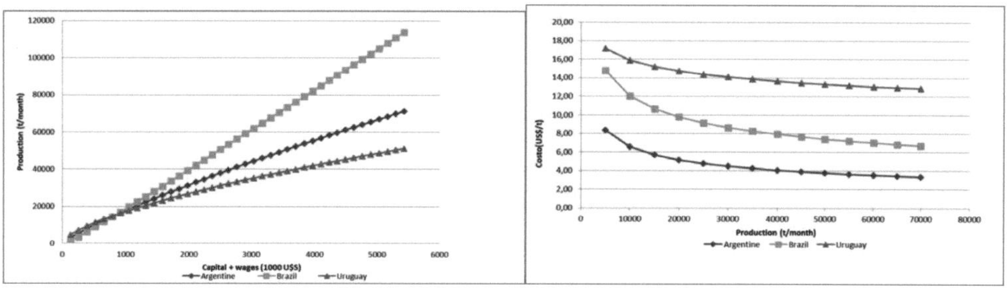

Figure 2.
Logarithm production and cost by ton models for Argentina, Brazil, and Uruguay.

The logarithmic production and cost functions by country from Mac Donagh et al. [11] are shown in **Figure 2**. Production per month increased with varying returns to scale as a function of capital and wages, and average total costs as a function of production per month were asymptotic L-shaped curves that roughly bisected the average costs listed in **Table 1**. The average total cost for Argentina had a minimum value at 18,000 tons per month with a cost of US$ 5.82 per ton. The Brazilian logging companies had higher production levels, at a somewhat higher level than Argentina. The lowest cost was of US$ 11.23 per ton, at a production of 25,000 tons per month. If production doubled, (50.000 tons per month) the cost decreased by 14%.

The results showed that that Brazilian loggers had the largest price elasticity to production changes, followed by Argentinean and Uruguayan logging firms. The elasticity of capital was 0.59 for Argentine loggers, 0.55 for Uruguay, and 0.4 in Brazil. For wages, the elasticity was 0.09 for Uruguay, 0.22 for Argentine and 0.63 for Brazilian loggers. Thus Brazil logging companies had the largest total factor elasticity (1.03), Argentina the second (0.81), and Uruguay the third (0.64). This indicates that at the mean values per country, additional capital would be most effective in Argentina (which has the least capital investment to date), and additional wages most effective in Brazil (which has the most capital).

2.2 Equipment technology and systems

Harvesting technologies have varied a lot in the period under consideration. While in the SE of the United States the most common mechanized systems are with feller-buncher and skidder, in the Southern Cone, there are different situations (**Table 2**). Feller-bunchers dominate in Argentina while they are very rare in Uruguay. Rubber tired harvesters are more frequent in Uruguay, while they are very rare in Brazil. In Uruguay, the harvester/forwarder system that is common in Scandinavia has been widely adopted, basically for eucalyptus clear cuttings. In Argentina, for clear pine felling, the most common is a system uses a feller-buncher, both crawled and wheel; rubber-tired skidders, and then processors at roadside. In pine thinning, although there are still operations with chainsaws, the most frequent are both small wheel harvesters, as well as small processors.

On the other hand, if we analyze the size of the companies, or the dispersion among small, medium and large companies, the Southern Cone had companies that produced from 5000 tons per month up to more than 100,000 tons per month. As noted, this generated average total logging costs per ton that varied from US$ 3.27 to US$ 25.81 per ton, with an average value of US$ 7.14 per ton for Argentina, US$ 8.41 per ton for Brazil, and US$ 14.16 per ton for Uruguay.

	Argentina (%)	Brazil (%)	Uruguay (%)	Total
Felling				
Harvesters	35	5	60	40
Fellers	48	39	13	23
Processors	20	57	23	91
Logging				
Forwarders	17	25	58	48
Skidders	47	45	8	38
Farm tractors	51	20	30	81
Loading	36	56	8	77

Source: Mac Donagh et al. [11].

Table 2.
Forest operations mechanization in Argentina, Brazil and Uruguay.

2.3 U.S. South production and cost comparisons

In the early literature for the Southeast U.S, smaller companies had computed average costs in the order of US $20 per ton [16], with companies that produced from 184 tons per month to 2600 tons per month for different harvesting system technology classes. More recently, Baker and Greene [17] report an increase in the size of logging contractors between 1987 and 2007 in Georgia. Companies with feller-skidder technology had reached an average production of 5828 tons per month per crew, or about 70,000 tons per year [18–21], with an investment of US$ 473,800 per crew [17].

In a more recent 2015 logger survey in North Carolina Coastal Plain, Hahn [2] found that loggers in eastern North Carolina produced 2960 tons of roundwood per firm per week, and worked 49 weeks per year. Thus the current North Carolina logging firm average production was 148,000 tons per year, although this did consist of some firms with multiple crews. Baker et al. [22] interviewed 22 logging firms across the South and reported average weekly production among those firms to be 4197 tons, which would be about 200,000 tons per year.

These average U.S. South production rates were usually on a per crew basis, but still were considerably less than the average rates of about 300,000 tons per year per firm in Brazil and Uruguay in 2012, which included some firms with multiple crews, and almost all of whom were harvesting quite uniform planted eucalyptus or U.S. Southern pine, mostly loblolly (*P. taeda*). Some of the higher reported individual firm average production rates in the Southern Cone were more than 400,000 tons per year. This might reflect the fact that loggers had multiple machines per crew—e.g., two feller-bunchers and several skidders—while the U.S system commonly only had one feller and two skidders. Several firms also ran multiple shifts per day, increasing production rates.

In addition, final harvest volumes in the Southern Cone were often almost twice as high, because the rapid growth per year and stocking per unit of area was that much greater than in the U.S. South [10]. With slightly shorter 18–22 year rotations, loblolly pine stands in the Southern Cone would produce and harvest more than 500 tons per hectare (200 tons per acre) at age 18–22, versus 250 tons per ha (100 tons per acre) in the U.S. South at age 25. This allowed much greater logging production from dense, closely spaced planted stands. Eucalyptus stand volumes were more like the 250 tons per ha (at ages 6–12), but the stands were almost all very similar clones in uniform stands, usually planted on relatively accessible lands, again facilitating timber harvesting.

Timber Mart-South [23, 24] showed a trend of increasing harvesting costs in the US Southeast. This variation goes from US $ 11 per ton 2009, to US $ 12.92 per ton for 2013. Baker et al. [25] estimated a logging cost index for the South, and set the initial cut and load at roadside logging value in 2011 as US$ 12.50 per ton. These values are somewhat greater than those that we calculated for Argentina and Brazil in the Southern Cone for the same period, but not Uruguay. Siry et al. [3], in a benchmarking comparison between different countries, found that logging costs for pulpwood was about US$ 11–13 per ton in the U.S. South, and US$ 5–7 per ton for Brazil. According Siry et al., low costs in Brazil were based on cutting-edge machinery, use of two to three shifts, and operators with a high training level.

3. North Carolina logging production and capacity

The timber harvesting literature reviewed above compares timber harvesting production and costs in the U.S. South with those of the Southern Cone at the beginning of the 2010s. From 2014 to 2016, we conducted research in North Carolina on timber harvesting production, costs, and capacity that provides an excellent benchmark for comparison to the prior Southern Cone research and earlier Southern U.S. literature. That research examined the status of logging and wood procurement in North Carolina, and whether we would have adequate harvesting capacity as timber production as mills increased their production after the recession of 2007. That research is summarized here, drawing from master's degree research by Hahn [2] and Roll [26] in cooperation with Cubbage.

Hahn [2] surveyed 27 procurement and logging firms in North Carolina, including questions about production, costs, and business environment. Roll [26] built on the results found by Hahn and focused on the question of a sufficient logging workforce, and derived estimates of logging production rates using a simulation approach based on "Arena" software, developed by Rockwell Automation [27]. In concert then, our research team developed a model of the timber supply chain path from the stump to the processing mills to estimate production levels of and requirements for harvesting crews in the Coastal Plains of North Carolina. These integrated efforts provide further insights into timber harvesting innovation, management, and capacity.

3.1 Logging capacity measurement

Adequate capacity for harvesting timber has been a continual issue for decades in the forestry sector. We analyzed timber harvesting and logging capacity in North Carolina's Coastal Plain region given structural economic trends and conditions after the recent 2007 recession and consequent downturn in the forest products industry [28]. Forest products industry profits have been squeezed by the housing and general economic recession, which has been forced down the supply chain to logging firms. Manufacturing capacity at pulp mills and sawmills has begun to return to pre-recession levels, but concerns remain that a reduced logging force could hinder the sustainability of manufacturing facilities. For reference, Baker et al. [25] report the median share of input costs for timber cut and haul average costs in 2011, with labor costs being the largest at 33% of the total. These are followed by fuel and oil (23%); equipment depreciation and interest (20%), repair and maintenance (11%), insurance (5%), and administration (4%).

Greene et al. [19–21] conclude that labor issues in the logging sector are a concern with an aging labor force and highly qualified workers seeking other employment, and note that the age of logging firm owners in the U.S. South has increased nearly 10 years over the past two decades. Not only is the workforce aging or altogether leaving, a similar survey reports that recruiting workers is also a significant concern to business success by firm owners [29]. A technical release from Wood Supply Research Institute [30] states that "new entrants into the wood supply business will be spotty at best," leading to the conclusion that any expansion in capacity with mostly be supported by existing wood suppliers. These discouraging trends in the U.S. might exacerbate any competitive disadvantages that were already evident as summarized in previous research.

Traditionally, the metric of "logging capacity" is expressed as how much a harvesting crew actually produces as a proportion of how much a crew could produce, during a given time period. While expressing logging capacity this way is certainly useful, it may not be the most pragmatic metric when assessing if the logging industry is capable of supplying adequate levels of wood fiber. Because most crews run at near-full mechanization, expansion of timber output is likely to depend most on scarce additional skilled labor to operate the increasingly sophisticated harvesting machinery. Thus our North Carolina study analyzed logging capacity in terms of additional labor required to meet wood demand.

3.2 North Carolina research methods

Face to face interviews were conducted between May and August 2014 with 27 subjects. Potential subjects' contact information was obtained from the North Carolina Forest Service website which contained a list of timber buyers, wood dealers, loggers, and mills in North Carolina. Subjects were contacted by telephone and notified of the scope and purpose of the interview. Of the 27 subjects, 13 were procurement foresters, six were wood dealers and eight were loggers. The subjects represented 23 different counties, with 21 in North Carolina, one in Virginia and one in South Carolina. Data from the surveys were summarized, and divided into regions—the western part of North Carolina with more relief and Mountains and hardwood species production; and the eastern part of the state, including the Piedmont and the Coastal Plain, and more pine timber production.

Subsequently, using the Arena Simulation software, we constructed a simulation model to approximate annual softwood and hardwood harvest levels for an individual logging crew. Next, we converted the most recent Timber Product Output (TPO) data in 2011 from cubic feet to tons using Timber Mart-South's weight equivalents to better understand historical harvest levels in our target market. Afterwards, employment in the logging industry (NAICS 113310) was retrieved from the Bureau of Labor Service (BLS) for Coastal Plain counties. Because the BLS reports employment as all employees paid by a logging firm, we consulted a previous logging survey in the coastal plain of Virginia to get an estimate of average "in-woods" crew sizes [31]. With this information we were able to estimate the number of logging crews required to produce 2011 levels of timber output as well as increased wood demand in subsequent years as forecasted by the Southern Forest Futures Project [32].

The simulation model designed in this study was subdivided into four interrelated segments in order to group activities that occur in the same harvesting activity (felling, skidding, sorting, loading). Each segment was simulated separately, and then linked into one integrated harvest simulation model using the Arena software package.

3.3 North Carolina results

3.3.1 Logging and procurement survey

Several results from the survey of procurement and logging firms [2] bear on the questions of timber harvesting production and capacity and comparisons with prior U.S. literature, as well as with the Southern Cone. The 13 procurement foresters and 6 wood dealer interviews indicated that for wood costs delivered to a mill, stumpage was the largest cost at 39%, logging costs were second at 35%, followed by hauling and wood dealer fees at 18% and 8% respectively (**Table 3**).

The eight logging firms interviewed had a mix of one single crew to several crews. In terms of operations, the average number of employees for each subject was 16, with a minimum of 2 and maximum of 30. Employee wages averaged at $13.75/hour, with the minimum being $11.00/hour and maximum of $15.50/hour. The average equipment spread was three feller-bunchers, three skidders, three loaders, one bulldozer, two chippers, seven chip vans, four trucks, five log trailers, and two processors.

The average length of time logging was 36 years with a minimum of 10 years and a maximum of 60 years. Average years that subjects had owned their business was 33 years with a minimum of 10 years and maximum of 60 years. Average weeks worked per year was 50 with a minimum of 47 and maximum of 52. Average hours worked per week was 51 with a minimum of 40 hours and maximum of 75 hours per week (**Table 4**). When asked to describe the type of operation, 62% subjects described their operation as an independent logging company. The remaining 37% were classified as contract loggers for wood dealers.

Average weekly production per firm was 1554 tons in pine and 1279 tons in hardwood (about 6200 and 5100 tons per month, or 76,000 and 63,000 tons per year respectively). For softwood product mix, the average breakdown was 47% pine pulpwood followed by 32% pine sawtimber, 12% pine chip-n-saw, and 9% pine biomass. For hardwood products, pulpwood was the most common product harvested at 44%, followed by sawtimber (32%), biomass (19%), and chip-n-saw (5%) respectively.

Category	North Carolina (%)	Western NC (%)	Eastern NC (%)
Stumpage	39	40	39
Logging	35	41	30
Hauling	18	19	16
Wood Dealer	8	0	15

Table 3.
Breakdown of average total delivered wood costs by category in North Carolina, 2014.

Business characteristic	Average	Minimum	Max
Years spent logging (years)	36	10	60
Years owned own business (years)	33	10	60
Weeks worked/year	50	47	52
Hours worked/week	51	40	75
Number of employees	16	2	30
Employee wage ($/hour)	$13.75	$11.00	$15.50

Table 4.
Business characteristics of loggers surveyed in North Carolina.

Logging rate or characteristic	North Carolina mean	North Carolina std. dev.	Western NC mean	Western NC std. dev.	Eastern NC mean	Eastern NC std. dev.
Cut and haul ($/ton)	13.74	1.65	15	2.85	13.33	1.18
Haul to mill ($/ton)	4.22	0.92	3.5	0.21	4.46	0.95
Haul distance (miles)	47	11	38	4	50	11

Table 5.
Logging payment rates in North Carolina, 2014.

Eastern logging crews were larger with an average of 18 employees, while western logging crews had an average of three employees. While there was no significant difference in the length of time each subject had been logging, there was a difference in the percentage that bought stumpage. In the western part of the state 100% of the subjects bought stumpage compared to 67% of subjects in the eastern region of the state. Likewise, 100% of western loggers were independent operations compared to 50% of eastern loggers.

The average timber harvesting cut and load rates were $13.74/ton and ranged from $12.00 to $17.00/ton. The average haul rate was $4.22/ton with a minimum of $3.35 and $5.50/ton. The average haul distance one way was 47 miles and ranged from 35 to 60 miles (**Table 5**).

3.3.2 Logging capacity simulation

The logging capacity simulation using the Arena software estimated the baseline scenario of the amount of timber harvesting production per crew in Eastern North Carolina as a case study, and then estimated how much added logging capacity would be required if additional product demand were increased from the low point in 2008 during the recession [26]. The Arena simulation estimated a logging operation of one feller-buncher, two skidders, and one loader as the typical harvesting spread based on the interviews from Hahn and other literature. Productivity rates by machine also were obtained from prior literature. These machine productivity rates were then entered into the simulation to model total harvesting system productivity per ton. The simulation results also were checked for validation with a spreadsheet with a generic harvesting tract based on the Auburn Harvesting Analyzer [33].

On average, each ton of wood spent 1.81 hours (108.6 minutes) in the entire harvest system from stump to loaded on a truck in the Arena simulations. Arena also provided instantaneous utilizations for all simulated harvesting machines. In the simulation, the feller-buncher was busy 54% of the time, one skidder was busy 78% of the time while the second was busy 69% of the time, and the loader was busy 23% of the time. Utilization rates in the Auburn Harvest Analyzer model were 54% for felling—the same as our simulation. AHA's utilization rate for the skidding function was 70%, and the skidding resources used in the simulation were busy on average a close 73.5% of the time. The loading procedure was utilized 26% of the time according to AHA's model, again close to our 23%. These utilization differences are minor and support the processing times programmed into each harvesting operation used in the simulation.

The simulation was run for two time periods to authenticate short-term and long-term production levels. To define a week in terms of hours worked, we adapted Hahn's [2] logging survey respondents in eastern North Carolina, and used an average of 49 hours worked per week. According to the survey, logging firms in eastern North Carolina produce 2960 tons of roundwood per week. After running

			COASTAL PLAIN TOTAL OUTPUT							
			**From TPO data & TMS Weight Equivalents							
	in Cubic Feet									
	All products		Sawlogs		Veneer logs		Pulpwood		Other industrial	
	Pine	Hard	Pine	Hard	Pine	Hard	Pine	Hard	Pine	Hard
All counties	**313,118,000**	**54,145,000**	**125,012,000**	**14,433,000**	**27,115,000**	**5,810,000**	**146,942,000**	**32,762,000**	**14,049,000**	**1,140,000**
	Cubic Meters (m^3)									
	All products		Sawlogs		Veneer logs		Pulpwood		Other industrial	
	Pine	Hard	Pine	Hard	Pine	Hard	Pine	Hard	Pine	Hard
All counties	**8,866,431**	**1,533,201**	**3,539,912**	**408,693**	**767,804**	**164,519**	**4,160,895**	**927,708**	**397,820**	**32,281**
	Short Tons (2000 lbs)									
	All products		Sawlogs		Veneer logs		Pulpwood		Other industrial	
	Pine	Hard	Pine	Hard	Pine	Hard	Pine	Hard	Pine	Hard
All counties	**10,786,412**	**2,241,439**	**4,306,463**	**582,836**	**934,068**	**231,580**	**5,061,916**	**1,382,549**	**483,965**	**44,475**
Other industrial: Includes composite panels, poles, posts, mulch, log homes, industrial fuelwood, and all other industrial products.										
Total Pine & Hard (tons)	**13,027,851**									

Table 6.
Timber product output weight equivalents, coastal plains total (tons).

the simulation for 49 hours, 2810 tons were produced, again a minor difference, of 150 tons per week. In terms of softwood and hardwood production, 2136 tons of softwood are produced and 674 tons of hardwood are produced.

According to Hahn's [2] survey, a logging crew in eastern North Carolina produces 147,975 tons per year. When we ran the simulation for 2450 hours, the harvesting model produces an annual total of 147,097 tons, a minor difference of 878 tons per year. The simulation model produced 111,793 tons of softwood and 35,304 tons of hardwood on an annual basis.

These production levels along with historical timber output reports, harvest distributions, employment metrics for the logging industry, and market forecasts for timber demand growth were used to analyze logging capacity.

Total timber output in both the USDA Forest Service Forest Inventory and Analysis (FIA) coastal plain regions in 2011 was 13,027,851 pine and hardwood tons, with about 83% of all harvests were pine according to the Forest Service Timber Product Output (TPO) report and about 76% of harvests were pine according to the Forest Service on-line annual timber removals package (FIDO), as summarized in **Table 6**.

Logging employment levels for North Carolina's forest products industry were obtained from tables of the Bureau of Labor Service [34]. In the fourth quarter of 2014, employment in the northern coastal plains' logging industry (NAICS 113310) totaled to 380, up from 236 in 2012. Conversely, logging employment in the southern coastal plain (NAICS 113310) was 344 during the fourth quarter of 2014, down from 372 in the first quarter of 2012. Labor force totals for the logging industry in the coastal plains equal 724, up from 608 in the first quarter of 2012.

Logging crew compositions vary among geographic regions. To gain an estimate of "in-woods" crew sizes, we consulted previous studies and logging surveys. According to Hahn's [2] logging survey, the average number of total employees per logging firm in North Carolina is 16 with a minimum of 2 and a maximum of 30, but this included multiple crews per firm. Baker et al. [22] found average crew sizes to be 4 and 4.2 employees via surveys and face-to-face interviews respectively. Finally, Barrett et al. [31] surveyed logging operations in Virginia's coastal plain and found average crew sizes to be 4.2 employees. Thus for the Coastal Plain, we chose to use an average crew size of 4.2 employees. Further, at the average 2011 Coastal Plain timber output levels of 13,027,851 pine and hardwood tons, and our simulation results of 147,097 tons to represent annual production levels of a typical logging crew in eastern North Carolina, then in 2011 it would have taken approximately 89 crews to supply TPO levels of wood. At 4.2 employees per crew, we estimated that 372 in-woods logging employees could supply the total production output for the North Carolina Coastal Plains in 2011.

Based on timber supply and demand projections from The Southern Forest Futures Project [32], three linear growth scenarios were developed and applied to analyze the sensitivity of the logging workforce to different rates of growth. The first scenario represented an additional 1.2-million-ton growth annually in timber output, or 0.5%. In scenario two, timber output increased by 1.289 million tons annually, or 1.0%. The third scenario increased at 1.5% per year, or 1.378 million additional tons annually. Under a 0.5% growth rate in timber output, an average of 8.2 additional logging crews or 34 logging employees will be needed each year to produce this level of output. At a 1.0% growth rate, an average of 8.77 additional crews per or 37 loggers will be needed each year to meet growth in production. Under a 1.5% growth rate in timber production, an additional 9.37 crews or 39 logging employees on average will be required each year over the 10-year period. Finally, to provide a "loggers required per million tons" metric, we estimate that an additional 29 "in-woods" employees (6.8 crews) are required each year to produce an additional 1 million green tons.

4. Discussion and conclusions

This chapter synthesizes several timber harvesting research threads and principles in the Southern Cone of South America and in North Carolina in the U.S. South, based on detailed empirical survey, economic, and simulation approaches. The research summarized here estimates timber harvesting productivity and costs in Southern Cone and in North Carolina; reviews the current literature about Southern U.S. timber harvesting productivity and costs; and provides unique components on innovation in the Southern Cone and on timber harvesting capacity in North Carolina. As such, the chapter provides a handy combination of current timber harvesting research, theory, and applications that can be useful for comparisons of comparative advantage between the U.S. South and the Southern Cone of South America, and provide benchmarking for logging and forest products firms, further research, or for policy considerations.

Some generalizations can be made about each of the possible international comparisons. Logging production data were calculated by each of our studies, and widely available in the literature. Cost data are reported in the Southern Cone research and the North Carolina survey, and some is available in the literature for both broad regions. The innovation research such as in our work in the Southern Cone is less common, but some comparisons can be made with other literature, and some insights can be drawn from our North Carolina research. Logging capacity has been talked about quite a bit, but our study in North Carolina is the only empirical example we found.

First, timber harvesting production rates have continued to increase along with mechanization for decades [35]. Carter et al. [16] reported the highest mechanized feller-buncher grapple-skidder system had average logging production rates of 2600 tons per month, or about 30,000 tons per year in 1990 in the U.S. South. By 2007, Baker and Greene [17] reported average production rates in Georgia of 5800 tons per month, or about 70,000 tons per year. By 2014, Hahn [2] found that logging firms in eastern North Carolina produce 2960 tons of roundwood per week, and work 49 weeks per year. Thus the North Carolina average firm production was 145,000 tons per year or 12,000 tons per month. Baker et al. [22] reported average weekly production among 22 firms in the South to be 4197 tons, which would be about 200,000 tons per year or 17,000 per month.

These average U.S. South production rates were usually on a per crew basis, but still were considerably less than the average rates of more than 300,000 tons per year per firm in Brazil and Uruguay in 2012. This is somewhat surprising that the Southern Cone is more productive, but reasonable upon reflection. Perhaps the biggest driver for this is that almost all timber harvesting in the Southern Cone is occurring in planted Southern U.S. pine species (e.g., *Pinus taeda*) or eucalyptus stands, which have much higher growth rates and much higher stand volumes per unit of area at harvest than in the planted and natural stands in the U.S. South—perhaps two to three times more volume per area at final harvest [10]. At least half of U.S. South timber production and logging still occurs in natural stands, which are less uniform and have lower stand volumes at harvest. Thus it would be easier to achieve high production rates with more wood volume per turn for machines and for systems. The reported production rates in South America also may include some multiple crews and multiple shifts per firm, which is not the case in most of the U.S. data.

As the productivity rates would suggest, timber harvesting costs per ton for the Southern Cone were usually less than in the U.S. South. For their logging cost index in the U.S. South, Baker et al. [25] set the initial cut and load at roadside logging value in 2011 as US$ 12.50 per ton. Hahn [2] found that reported cut and load logging contract rates in North Carolina (which presumably include some profit, so should be higher than logging costs) were US$ 13.74 per ton. These values were somewhat greater than those found in most of the production in the Southern Cone for about the same period—US$ 7.41 per ton in Argentina, US$ 8.19 per ton in Brazil—although less than the $14.82 per ton in Uruguay.

The rank of this difference makes sense since productivity rates are higher in the Southern Cone, if the equipment fixed and operating costs are similar in all countries. Uruguay costs probably were somewhat greater because they did use more expensive full timber processor/forwarder equipment, and had high fuel costs as well. The higher timber production volumes per area at harvest and the use of multiple shifts help drive these higher logging productivities and lower logging costs. While we do not review logging transport costs here, they too are problematic in both the U.S. South, where it is hard to find drivers who meet the strict license requirements, and in the southern Cone, where transportation networks often are not very good.

Mac Donagh [15] analyzed the role of innovation specifically in the performance of logging firms in the Southern Cone. That study showed that that the mechanization and innovation process there has been through the diffusion and adoption process throughout much of plantation forestry sector in the Southern Cone, and that good management will lead to better outcomes and more profits there like developed countries in the northern hemisphere. Loggers that had mutualistic, cooperative relations with contracting pulp and paper companies had the highest production rates and the lowest production cost per ton—which was the opposite of findings in more developed countries. Mechanization with the highest technology produced the highest production output levels.

Logging innovation capacities were more important than the business skills for production and growth. Innovation was more important than just buying the newest equipment in determining firm success [15]. In both the Southern Cone and in the study by Stone et al. [14] in Maine, the research found that the most successful logging contractors were the best innovators in products or process, followed by market innovation.

The last research focus on timber harvesting and logging capacity in this chapter was only directly investigated by Roll [26], although we can infer much about this question from other research and popular articles. In brief, there are pervasive concerns that as the forest products manufacturing sector expands, at least in the U.S., there will not be enough loggers to harvest all the wood needed [36]. Timber harvesting is hard, dangerous, and the pay is relatively modest. Surveys by Hahn [2] and Baker and Greene [17] and Greene et al., [19–21] indicate that both the average age of loggers, and often the age of their equipment, is getting much older.

While the logger demographics are disconcerting, there were high production levels in North Carolina as found by Hahn [2], and corroborated by the simulations of Roll [26] and others in the South. Roll determined that a relatively modest number of 30 employees and seven harvesting crews of four in-woods persons could harvest an additional 1 million tons of wood in the Coastal Plain of North Carolina. However, this is only an 8% increase based on 13 million tons of production in the state; large production increases could require many more workers quickly.

We conclude that an expansion of the number of logging crews—which can harvest from 150,000 to 200,00 tons of production per year—would most likely come from expansion and innovation from existing logging firms, not completely new entrants to the business, or from wood dealers who serve as middlemen in the procurement process in Eastern North Carolina. Thus an expansion of moderate amounts of timber harvesting production at the margin does not seem insurmountable, although several million tons or more of production would become increasingly harder to achieve by merely expanding current, often undercapitalized, logging firms. Innovation to achieve higher productivity would be important in this case.

While not examined specifically, we believe it would be easier for logging companies in the Southern Cone to expand to achieve higher logging production to supply forest products manufacturing facilities. The logging production rates exceeding 300,000 tons per year were higher than in the U.S. South; the work is still perceived comparatively favorably by rural workers in the Southern Cone; and there are fewer rural manufacturing or service alternatives to timber extraction.

Overall, these linked studies of timber harvesting/logging in the Southern Cone of South America and the U.S. South are very informative. While the timber harvesting productivity rates and costs were not far apart, the Southern Cone generally had a competitive advantage, with observably higher logging production rates and lower logging costs. Timber harvesting technology and innovation had matured considerably in the Southern Cone in the last two decades. The studies reviewed here suggest that average logging cut and haul rates were perhaps \$3–\$5 per ton cheaper in the Southern Cone, which multiplied by one quarter to a million tons per sawmill or pulp mill, adds up to a considerable cost advantage per mill, and for South America.

In addition, the timber plantation growth rates, final harvest yields, and investment returns for stumpage alone also are much better in South America than the U.S. South [10]. Cheaper fast grown timber plantation costs and timber harvesting production and cost advantages unite to provide substantial competitive advantages throughout the value chain to Southern Cone planation forestry. These countries of course have considerable challenges and more variable macroeconomic factors, political risk, poor roads and infrastructure, and other issues, which constrain their excellent forestry opportunities.

However, if the underlying institutional fundamentals do align well, the Southern Cone can grow and harvest wood cheaply, and will continue to expand forest products mills and harvesting capacity more quickly as well. This trend is evidenced by the opening of many new pulp and paper mills in the Southern Cone over the last few decades, while a large number have closed in the U.S. South. The Southern Cone roundwood production increased about 25% from 171 million m^3 in 2008 to 217 million m^3 in 2017 [5, 6], while that in the U.S. South was relatively flat at about 185 million m^3 [7]. The South still has the most forest manufacturing production capacity in the world, but has a decreasing total output share, which is likely to continue based on the timber harvesting costs examined here, as well as timber plantation production and cost advantages in the Southern Cone.

Acknowledgements

Partial funding for the research on was provided by the USDA Forest Service Southern Research Station; by the NC State University Southern Forest Resource Assessment Consortium; and by the respective organizations of the co-authors. Karen Abt of the USDA Forest provided insightful review comments for the manuscript.

Author details

Patricio Mac Donagh[1], Joshua Roll[2], George Hahn[3] and Frederick Cubbage[4*]

1 Universidad Nacional de Misiones, Eldorado, Misiones, Argentina

2 International Paper-Mansfield & Campti Mills, LA, USA

3 Virginia Polytechnic Institute and State University, Blacksburg, Virginia, USA

4 North Carolina State University, Raleigh, North Carolina, USA

*Address all correspondence to: fredcubbage@yahoo.com

References

[1] Fricko O, Havlik P, Rogelj J, Klimont Z, Gusti M, Johnson N, et al. The marker quantification of the shared socioeconomic pathway 2: A middle-of-the-road scenario for the 21st century. Global Environmental Change. 2016;**42**(1):251-267. DOI: 10.1016/j.gloenvcha.2016.06.004

[2] Hahn G. Wood procurement and harvesting trends in North Carolina. A report submitted to the Graduate Faculty of North Carolina State University in partial fulfillment of the requirements for the Degree of Master of Forestry. 2015. 25 p. Mimeo. https://repository.lib.ncsu.edu/bitstream/handle/1840.4/8618/Hahn%2C%20George%20final.pdf?sequence=1&isAllowed=y

[3] Siry JP, Greene WD, Harris TG Jr, Izlar RL, Hamsley AK, Eason K, et al. Wood supply chain efficiency and fiber cost: What can we do better? Forest Products Journal. 2006;**56**(10):4-10

[4] Sedjo RA. The comparative economics of plantation forestry: A global assessment. Washington, D.C.: RFF Press, Routeldge Ed; 1983

[5] FAO. State of the World's Forests 2011. Food and Agriculture Organization of the United Nations. Rome, Italy; 2011

[6] FAO. FAOStat. Roundwood production by country, 2017. 10 February 2019. Accessed at: http://www.fao.org/faostat/en/#data/FO

[7] USDA Forest Service. Timber Products Output (TPO) Studies. 2015. Accessed at: https://www.fia.fs.fed.us/program-features/tpo/

[8] Conrad JL IV, Greene WD, Hiesl P. A review of changes in US logging businesses 1980s–present. Journal of Forestry. 2018;**116**(3):291-303

[9] Gonzalez RW, Saloni D, Sudipta D, Cubbage F. South America: Industrial roundwood supply potential. BioResources. 2008;**3**(1):255-269

[10] Cubbage F, Mac Donagh P, Balmelli G, Morales Olmos V, Bussoni A, Rubilar R, et al. Global timber investments and trends, 2005-2011. New Zealand Journal of Forestry Science. 2014;**44**(Suppl 1):57. Accessed at: http://www.nzforestryscience.com/content/4451/s7. [Accessed: December 5, 2014]

[11] Mac Donagh P, Botta G, Schlichter P, Cubbage F. Harvesting contractor production and costs in forest plantations of Argentina, Brazil, and Uruguay. International Journal of Forest Engineering. 2017;**28**(3):157-168. DOI: 10.1080/14942119.2017.1360657

[12] OECD and Eurostat. Guidelines for Collecting and Interpreting Innovation Data (Oslo Manual). Paris, France: OECD; 2016. p. 183

[13] Lindroos O, La Hera P, Häggström C. Drivers of advances in mechanized timber harvesting–A selective review. Croatia Journal of Forest Engineering. 2017;**38**(2):243-258

[14] Stone IJ, Benjamin JG, Leahy J. Applying innovation theory to Maine's logging industry. Journal of Forestry. 2011;**109**:462-469

[15] Mac Donagh PM. Los contratistas de cosecha de plantaciones forestales: Análisis de las funciones de producción y de su comportamiento como innovadores. Ph.D. Thesis. Universidad de Buenos Aires, Área Ciencias Agropecuarias;2016 183 p.

[16] Carter DR, Cubbage FW, Stokes BJ, Jakes PJ. Southern pulpwood harvesting productivity and cost changes between 1979 and 1987. USDA Forest Service Research Paper NC-318. 1994. p. 33

[17] Baker S, Greene WD. Changes in Georgia's logging workforce, 1987-2007. Southern Journal of Applied Forestry. 2008;**32**(2):60-68

[18] Bolding MC, Barrett SM, Munsell JF, Groover MC. Characteristics of Virginia's logging businesses in a changing timber market. Forest Products Journal. 2010;**60**(1):86-93

[19] Greene WD, Marchman SC, Baker SA. Changes in logging firm demographics and logging capacity in the US South. In: Proceedings of the 36th Annual Council on Forest Engineering Meeting. 2013a. Available from: https://secure.hosting.vt.edu/www.cofe.frec.vt.edu/documents/2013/Greene_Marchman_Baker.pdf

[20] Greene WD, Marchman SC, Baker SA. Changes in logging firm demographics and logging capacity in the U.S. South. In: Proceedings of the 36th Annual Council on Forest Engineering Meeting, July 7-10, 2013, Missoula, MT; 2013b. p. 7

[21] Greene WD, Marchman SC, Baker SA. Changes in demographics and logging capacity in the U.S. South. In: Proceedings of the 36th Annual Council on Forest Engineering Meeting. Missoula, MT. 2013c. 7 p. (Cited in Conrad et al. 2018, supra)

[22] Baker S, Greene D, Mei R, Langdale H. Verification of the UGA Logging Cost Index. Wood Supply Research Institute; 2015 https://wsri.org/resources/media/VerificationOfLCI-UGA-FinalReport.pdf

[23] Timber Mart-South. Timber Mart-South Logging Rates. Timber Mart-South. 2007. Available from http://www.timbermart-south.com/pdf/samplereport/LogRateOtherProdSample.pdf

[24] Timber Mart-South. 3rd Quarter. Abridged Market News Quarterly. Athens, GA: Warnell School of Natural Resources. University of Georgia. 2013. p. 11

[25] Baker SA, Mei B, Harris TG, Greene WD. An index for logging cost changes across the US South. Journal of Forestry. 2014;**112**(3):296-301

[26] Roll J. Assessment of timber harvesting and logging capacity in North Carolina's Coastal Plain—A simulation approach. M.S. Thesis. North Carolina State University; 2016. http://www.lib.ncsu.edu/resolver/1840.16/11148

[27] Rockwell International. 2016. Arena software. https://www.rockwellautomation.com/rockwellsoftware/simulation.page

[28] Woodall C, Ince P, Skog K, Aguilar F, Keegan C, Sorenson C, et al. An overview of the forest products sector downturn in the United States. Forest Products Journal. 2011;**61**(8):595

[29] Ward N. Attracting Workers to Logging Employment: Survey Report. Washington, DC: Forest Resources Association; 2013

[30] Taylor D, Barynin P. Is the Forest Industry Wood Supply Chain Ready for Recovery? Forest Resources Association, Inc. 2013. Available from http://wsri.org/resources/media/13r25.pdf

[31] Barrett SM, Bolding MC, Munsell JF. Characteristics of logging businesses across Virginia's diverse physiographic regions. Forests. 2017;**8**:468. Special Issue: Forest Operations, Engineering and Management. DOI: 10.3390/f8120468

[32] Wear D, Prestemon J, Hugget R, Carter D. Southern Forest Futures Project Technical Report 178-Markets; 2013. Available from http://srs.fs.usda.gov/futures/technical-report/#ch9

[33] AHA. 2007. Auburn Harvest Analyzer. Microsoft Excel Model. Adapted by W.D. Greene and F.W. Cubbage

[34] BLS. Quarterly Census of Employment and Wages-Characteristics of the Data-Employment (US). 2014. Bureau of Labor Service. Available from http://www.bls.gov/cew/cewbultn12.htm

[35] Stewart P. US South Wood Supply Trends 1995-2015. Forest2Market; 2015. Available from http://www.theusipa.org/Documents/USSouthWoodSupplyTrends.pdf

[36] Knight D. Nowhere to Turn. Timber Harvesting. May/June 2011: 10-19